00-CTS-16ET

ACCLIMATION OF NITRIFIERS FOR ACTIVATED SLUDGE TREATMENT: A BENCH-SCALE EVALUATION

by:

Robert A. Zimmerman

Andrew T. Bradshaw

Don Richard

City of Moorhead, Minnesota

2004

Water Environment Research Foundation

The Water Environment Research Foundation, a not-for-profit organization, funds and manages water quality research for its subscribers through a diverse public-private partnership between municipal utilities, corporations, academia, industry, and the federal government. WERF subscribers include municipal and regional water and wastewater utilities, industrial corporations, environmental engineering firms, and others that share a commitment to cost-effective water quality solutions. WERF is dedicated to advancing science and technology addressing water quality issues as they impact water resources, the atmosphere, the lands, and quality of life.

For more information, contact:
Water Environment Research Foundation
635 Slaters Lane, Suite 300
Alexandria, VA 22314-1177
Tel: (703) 684-2470
Fax: (703) 299-0742
www.werf.org
werf@werf.org

This report was co-published by the following organizations. For nonsubscriber sales information, contact:

IWA Publishing
Alliance House, 12 Caxton Street
London SW1H 0QS, United Kingdom
Tel: +44 (0) 20 7654 5500
Fax: +44 (0) 20 7654 5555
www.iwapublishing.com
publications@iwap.co.uk

Library of Congress Catalog Card Number: 2003115650

IWAP ISBN: 184339-691-2

City of Moorhead, Minnesota

ACKNOWLEDGMENTS

The project team acknowledges the City of Moorhead, Minnesota, for providing facilities and financial support for this project and thanks the staff of the Moorhead Wastewater Treatment Facility for their outstanding effort in completing the project.

Report Preparation

Principal Investigator:
Robert A. Zimmerman, Ph.D., P.E.
City of Moorhead, Minnesota

Project Team:
Andrew T. Bradshaw, BSES
Don Richard, Sc.D., P.E.
City of Moorhead, Minnesota

Project Subcommittee

H. David Stensel, Ph.D., P.E.
University of Washington, Seattle

Water Environment Research Foundation Staff

Director of Research: Daniel M. Woltering, Ph.D.
Director of Technical Services: Amit Pramanik, Ph.D.

ABSTRACT AND BENEFITS

Abstract:

Nitrification kinetics were evaluated in bench-scale batch reactors. These reactors were fed with a synthetic wastewater containing approximately 1000 mg ammonia-nitrogen (NH_3-N)/L operated at 5-, 10-, and 20-day solids retention times (SRTs), and with dewatered biosolids supernatant (1126–1680 mg NH_3-N/L) operated at a 20-day SRT.

The presence of un-ionized ammonia and un-ionized nitrous acid appeared to inhibit complete nitrification for the 5- and 10-day SRTs. Near-complete nitrification was observed for both substrates for the 20-day SRT. Observed ammonium oxidation rates decreased with increasing SRT. Observed yield coefficients were similar for all SRTs and substrates.

Higher SRTs showed fully established steady-state conditions despite process start-up and operational considerations. Although it may be possible to culture a nitrifier population capable of near-complete nitrification at lower SRTs, the design configuration and operational strategy must mitigate the potential for un-ionized ammonia and un-ionized nitrous acid inhibition (e.g., process start-up at lower concentration with gradual increase to higher concentration, continuous feed operation).

Batch bioaugmentation analyses were conducted in the mixed liquor suspended solids and final clarifier effluent. The effluent was from a non-nitrifying activated sludge with seed nitrifiers developed from the 20-day SRT reactors and with biomass from a nitrifying trickling filter facility. Ammonia removal in all observed bioaugmentation analyses had no apparent lag or acclimation period. Observed ammonium oxidation rates did not differ significantly between the seed and batch bioaugmentation reactors. Acclimation does not appear to be a critical obstacle for nitrifier bioaugmentation when environmental conditions (e.g., temperature, pH) between the seed and bioaugmentation processes are not significantly different.

Benefits:

- Suggests the potential for inhibition of complete nitrification by un-ionized ammonia and un-ionized nitrous acid in high-strength ammonia wastewater treatment.
- Identifies the need to consider inhibitory conditions in the design a full-scale high-strength ammonia nitrification system, particularly at lower SRTs.
- Emphasizes process start-up and operational considerations in the design of a full-scale high-strength ammonia nitrification system.
- Indicates that acclimation of separately cultured nitrifiers (from high-strength ammonia wastewater) is not a critical obstacle for bioaugmentation in non-nitrifying activated sludge under similar environmental conditions (e.g., temperature and pH).
- Provides a technical basis for continuous-flow pilot studies of activated sludge bioaugmentation with nitrifiers developed from high-strength ammonia wastewater.
- Indicates that conversion to nitrate was achieved at higher SRT values exhibiting steady-state operation.

Keywords: nitrification, bioaugmentation, inhibition, solids retention time

TABLE OF CONTENTS

LIST OF TABLES

LIST OF FIGURES

LIST OF ACRONYMS

μ_N	nitrifier specific growth rate
μ_N'	nitrifier net specific growth rate
μ_{Nmax}	nitrifier maximum specific growth rate
μ_{Nest}	nitrifier estimated specific growth rate
μ_{Nobs}	nitrifier observed specific growth rate
θ_c	solids retention time
APHA	American Public Health Association
b_N	nitrifier endogenous decay coefficient
COD	chemical oxygen demand
DO	dissolved oxygen
HNO_2-N	nitrous acid
K_N	half-saturation coefficient for nitrification
MLSS	mixed liquor suspended solids
N	NH_3-N concentration
N_0	initial NH_3-N concentration
N_1	final NH_3-N concentration
NH_3	ammonia
NH_3-N	ammonia-nitrogen
NH_4-N	ammonium nitrogen
NO_2-N	nitrite-nitrogen
NO_3-N	nitrate-nitrogen
q_N	ammonium oxidation rate
q_{Nmax}	maximum ammonium oxidation rate
q_{Nest}	estimated ammonium oxidation rate
q_{Nobs}	observed ammonium oxidation rate
R^2	coefficient of determination
SRT	solids retention time (θ_c)
S.U.	standard units
t	hydraulic detention time
TSS	total suspended solids
TVSS	total volatile suspended solids
U.S. EPA	United States Environmental Protection Agency
VSS	volatile suspended solids
VSS_N	active nitrifier concentration as VSS
X	VSS concentration
X_0	influent VSS concentration
X_N	active nitrifier concentration as VSS
X_{Nest}	estimated active nitrifier concentration as VSS
Y_N	yield coefficient for nitrification
Y_{Nest}	estimated yield coefficient for nitrification
Y_{Nobs}	observed yield coefficient for nitrification

Executive Summary

ES.1 Project Objectives

The design and operation of conventional activated sludge systems usually generate conditions that generally are not conducive to nitrification. Because the yield of heterotrophic organisms is greater than the yield of autotrophic nitrifying bacteria, the autotrophic population normally constitutes only a small fraction of the total biomass. Kinetic theory suggests a basis for activated sludge bioaugmentation with nitrifiers developed from high-strength ammonia wastewater such as dewatered biosolids supernatant. Application of this theory requires practical investigation of the conditions at which the side-stream (or seed) system must be operated to generate a supplemental nitrifier population, the quantity of nitrifiers that can be generated for bioaugmentation purposes, and the ability of supplemental nitrifiers to acclimate to new conditions when used for bioaugmentation. Because the nitrification process is sensitive to a variety of chemical inhibitors and environmental conditions, the potential for unfavorable conditions in the side-stream reactor and/or during bioaugmentation should be evaluated.

A fundamental requirement for successful bioaugmentation is the ability of the separately cultured nitrifiers to rapidly acclimate to new environmental conditions presented by the mixed culture activated sludge system. The acclimation period, if any, may affect the quantity of supplemental nitrifiers required.

The objective of this research was to investigate

- the potential to culture nitrifying microorganisms in a high-strength ammonia substrate using high-concentration start-up conditions and batch-feed operation, and
- the ability of these cultures to adapt or acclimate to new conditions when used to bioaugment a non-nitrifying activated sludge process operated at similar environmental conditions (e.g., temperature, pH) to achieve nitrification.

ES.2 Organization of Report

This report includes a brief review of selected literature related to the concept of bioaugmentation in conventional activated sludge systems using supplemental nitrifiers developed from high-strength ammonia wastewater (Chapter 1.0); a description of the experimental methods used to evaluate the development of supplemental nitrifier populations for subsequent bioaugmentation and the bioaugmentation procedures employed (Chapter 2.0); results from bench-scale studies of the development of nitrifying cultures in high-strength ammonia wastewater under high concentration start-up conditions and batch-feed operation (Chapter 3.0); results from bioaugmentation analyses in mixed liquor suspended solids (MLSS) and final clarifier effluent from a non-nitrifying activated sludge system operated at similar environmental conditions to the seed bioreactors (Chapter 4.0); and conclusions and recommendations for additional continuous-flow pilot studies of this concept (Chapter 5.0).

ES.3 Research Approach

This laboratory bench-scale investigation was conducted in two phases. The first phase evaluated nitrification kinetics in bench-scale batch reactors. The reactors were fed with synthetic wastewater containing approximately 1000 mg ammonia-nitrogen (NH_3-N)/L and operated at 5-, 10-, and 20-day solids retention times (SRTs). A fourth reactor was fed with dewatered biosolids supernatant (1126–1680 mg NH_3-N/L) and operated at a 20-day SRT. Process start-up used high-strength waste directly. Reactors were operated on a batch-feed (once per day) basis.

In the second phase, cultures from the 20-day SRT reactors were used in bench-scale bioaugmentation analyses to evaluate acclimation. Biomass obtained from a nitrifying trickling filter facility was also used for comparison. Bioaugmentation analyses were conducted in the MLSS and final clarifier effluent from a non-nitrifying activated sludge system. In all bioaugmentation analyses, the temperature and pH of the seed reactor and the bioaugmentation reactor were not significantly different so that external environmental stress conditions would not be inhibitory to the nitrification process.

ES.4 Conclusions

Partial conversion of ammonia to nitrite, with little or no conversion of nitrite to nitrate, occurred in the synthetic wastewater at a 5-day SRT. Un-ionized ammonia appeared to inhibit ammonium oxidation and prevent nitrite oxidation. At a 10-day SRT, ammonia removal, nitrite production and removal, and nitrate production were cyclical and appeared to be related to inhibitory concentrations of un-ionized ammonia and un-ionized nitrous acid. Fully established steady state conditions were not observed at the 5- and 10-day SRTs under the conditions evaluated (i.e., process start-up using high-strength ammonia wastewater and batch-feed operation).

Near-complete nitrification was observed for high-strength ammonia synthetic wastewater and dewatered biosolids supernatant at a 20-day SRT without inhibitory conditions despite process start-up and operational considerations affecting lower SRTs.

Kinetic parameters developed from average daily data are shown in Table ES-1. Observed ammonium oxidation rates (q_{Nobs}), yield coefficients (Y_{Nobs}), and specific growth rates (μ_{Nobs}) were computed on the basis of the total VSS concentration (X). Where possible, estimated ammonium oxidation rates (q_{Nest}), yield coefficients (Y_{Nest}), and specific growth rates (μ_{Nest}) were computed on the basis of the estimated active nitrifier VSS concentration (X_{Nest}).

In all of the seed bioreactors, complete nitrification appeared to be dependent on the concentrations of ammonia, un-ionized ammonia, and/or un-ionized nitrous acid. Estimated ammonium oxidation rates decreased with increasing SRT whereas estimated yield coefficients were similar for all SRTs. Stable nitrifying populations and fully established steady-state conditions were not observed at lower SRTs. Although it may be possible to culture a population capable of complete nitrification at a lower SRT, the design configuration and operational strategy must consider, mitigate, and/or eliminate the potential for un-ionized ammonia and un-ionized nitrous acid inhibition (e.g., process start-up at lower concentration with gradual increase to higher concentration, continuous feed operations).

Table ES-1. Kinetic Parameters for Bioreactor Performance from Average Daily Data.

Parameter	Synthetic waste			Supernatant
	5-day SRT	10-day SRT	20-day SRT	20-day SRT
Temperature, °C	20.0±2.0	20.2±1.9	20.2±1.8	20.2±2.6
pH, standard units (S.U.)	7.0 – 7.4	7.0 – 8.6	7.0 – 7.8	6.0 – 9.1
DO, mg/L	8.3±1.1	8.0±3.2	8.0±1.7	8.1±1.8
Influent NH_3-N, mg/L	1005	989	996	1440
Influent VSS, mg/L	9.0	13.2	13.5	327
Effluent NH_3-N, mg/L	736	190	15.7	1.1
Effluent VSS (X), mg/L	35.7	92.5	148	670
X_{Nest}, mg/L	35.7	92.5	148	343
Y_{Nobs}, g VSS/g NH_3-N removed	N/A	N/A	N/A	0.47
q_{Nobs}, g NH_3-N oxidized/g VSS per day	N/A	N/A	N/A	0.11
μ_{Nobs}, d^{-1}	N/A	N/A	N/A	0.05
Y_{Nest}, g VSS/g NH_3-N removed	0.13	0.12	0.15	0.24
q_{Nest}, g NH_3-N oxidized/g VSS per day	1.51	0.86	0.33	0.21
μ_{Nest}, d^{-1}	0.20	0.10	0.05	0.05

In Phase 2 of the study, batch bioaugmentation analyses were conducted in the MLSS and final clarifier effluent from a non-nitrifying activated sludge with seed nitrifiers developed from the high-strength ammonia synthetic wastewater and dewatered biosolids supernatant operated at a 20-day SRT and also with biomass from an operating nitrifying trickling filter facility. Environmental conditions (e.g., temperature and pH) between the seed bioreactors (used to develop the supplemental nitrifiers) and the bioaugmentation reactors (to which supplemental nitrifiers were added) were not significantly different. For nitrifiers developed from high-strength ammonia synthetic waste, the difference in temperature and pH between the seed and bioaugmentation reactors was ±2.1°C and ±0.5 standard units (S.U.), respectively. For nitrifiers developed from dewatered biosolids supernatant, the difference in temperature and pH between the seed and bioaugmentation reactors was ±4.5°C and ±0.8 S.U., respectively. Results of the bioaugmentation analyses are summarized in Tables ES-2 through ES-4.

Table ES-2. Bioaugmentation Analyses with Nitrifiers from High-strength Ammonia Synthetic Waste at 20-day SRT.

Run #	1	4	5
Substrate	Seed bioreactor	Clarified activated sludge	Activated sludge
Substrate volume, L	N/A	3.50	3.50
Supplemental nitrifier volume, L	N/A	0.50	0.50
Total volume, L	5.00	4.00	4.00
X_{Nest} (VSS_N), mg/L	109	14.2	15.0
NH_3-N removal rate, mg/L per hr	5.78	0.94	0.86
q_{Nest}, mg NH_3-N oxidized/mg VSS_N per day	1.28	1.59	1.38

Table ES-3. Bioaugmentation Analyses with Nitrifiers from Dewatered Biosolids Supernatant at 20-day SRT.

Run #	6	9	10	11
Substrate	Seed bioreactor	Clarified activated sludge		Activated sludge
Substrate volume, L	N/A	4.75	3.55	3.55
Supplemental nitrifier volume, L	N/A	0.25	0.45	0.45
Total volume, L	5.00	5.00	4.00	4.00
X (VSS), mg/L	558	31.3	61.4	77.2
X_{Nest} (VSS_N), mg/L	285	16.0	31.3	39.4
NH_3-N removal rate, mg/L per hr	6.55	0.64	0.72	0.65
q_{Nobs}, mg NH_3-N oxidized/mg VSS per day	0.28	0.49	0.28	0.20
q_{Nest}, mg NH_3-N oxidized/mg VSS_N per day	0.55	0.97	0.55	0.40

Ammonia removal was observed in all bioaugmentation analyses with no apparent lag or acclimation period. For nitrifiers developed from high-strength ammonia wastewater and dewatered biosolids supernatant, nitrification occurred sequentially with the conversion of ammonia to nitrite taking place before the conversion of nitrite to nitrate. Observed and estimated ammonium oxidation rates between seed and bioaugmentation reactors were also used as a criterion to evaluate acclimation. As indicated in Tables ES-2 and ES-3, the ammonium oxidation rates in the bioaugmentation analyses were not significantly different from the seed bioreactors. Therefore, acclimation does not appear to be a critical obstacle for nitrifier bioaugmentation in conventional activated sludge systems when environmental conditions between the seed and bioaugmentation processes are not significantly different.

Table ES-4. Bioaugmentation Analyses with Nitrifiers from Trickling Filter Biomass.

Run #	15	16	17 (duplicate)	18	19
Substrate	Clarified activated sludge			Activated sludge	
Substrate volume, L	5.730	5.864	5.685	5.767	5.883
Supplemental nitrifier volume, L	0.270	0.136	0.315	0.233	0.117
Total volume, L	6.000	6.000	6.000	6.000	6.000
X (VSS), mg/L	490	320	440	528	265
NH_3-N removal rate, mg/L per hr	1.71	1.11	1.24 (1.27)	1.46	0.81
q_{Nobs}, mg NH_3-N oxidized/mg VSS per day	0.084	0.083	0.067 (0.069)	0.067	0.074

Nitrate production occurred almost immediately with the trickling filter biomass. As shown in Table ES-4, observed ammonium oxidation rates for the bioaugmentation analyses with nitrifying trickling filter biomass were significantly less than those from the dewatered biosolids supernatant seed reactor indicating that the trickling filter biomass contained a lower fraction of active nitrifiers.

ES.5 Recommendations

It appears that there is sufficient technical basis to conduct additional study of this concept. A continuous-flow pilot study of the complete system is recommended. The pilot study should include a sidestream treatment system for developing nitrifiers from high-strength ammonia wastewater and a mainstream non-nitrifying activated sludge process. In particular, further study should consider the optimal sidestream system design, start-up, and operation at an appropriate SRT to provide a yield of nitrifiers sufficient to achieve complete nitrification in the mainstream activated sludge process.

CHAPTER 1.0

INTRODUCTION

1.1 Background

Conventional activated sludge systems are normally designed and operated at conditions that are generally not conducive to biological nutrient removal (nitrification/denitrification). In essentially all domestic and in most industrial wastewaters, the concentration of carbonaceous organic matter is much greater than the concentration of nitrogen (Sedlak, 1991). Because the yield of heterotrophic organisms is greater than the yield of autotrophic nitrifying bacteria, the autotrophic population normally constitutes only a small fraction of the total biomass. Therefore, to build and sustain a sufficient nitrifying population, a minimum solids retention time (SRT) that is based on nitrification kinetics must be maintained. As a result of the slow growth rate of nitrifiers, this SRT is much greater than that required for heterotrophic bacteria. The higher SRT normally required to accomplish nitrification (10–15 days) generally cannot be achieved in a conventional activated sludge system (U.S. EPA, 1993). Slow-growing nitrifiers are, therefore, typically "washed out" of conventional activated sludge systems.

Nitrification is very temperature dependent (U.S. EPA, 1993). To provide an inexpensive alternative for treatment facilities in northern climates to upgrade for cold weather nitrification, In 1998 Kos proposed the concept of a short SRT vitrification process/flowsheet. Although this concept was initially developed for cold weather nitrification, it may be expanded for nitrification in non-nitrifying conventional activated sludge systems in general.

In the short SRT treatment scheme, a portion of a solids dewatering supernatant side-stream (typically with a high ammonia concentration) is isolated and used to culture nitrifying microorganisms. Although the use of an internal high-strength ammonia side-stream may be beneficial in culturing nitrifiers, it is not a necessity. Supplemental nitrifiers may be obtained from alternative substrates for the same purposes. The biological solids produced in the side-stream reactor should contain a high percentage of nitrifiers in addition to other generally inert particles from the biosolids supernatant. These nitrifiers could subsequently be introduced directly into the mainstream activated sludge system. The quantity of supplemental nitrifiers introduced would need to sufficient to continuously replace those that are washed out (removed with waste activated sludge) at shorter SRTs.

Based on theoretical background and kinetic modeling, the addition of supplemental nitrifiers would allow the mainstream activated sludge process to nitrify at very low SRTs (Kos, 1998). The higher SRT required to prevent nitrifier washout is offset by the continual addition of nitrifiers. Based on this theory, nitrification should be achieved at SRTs of 7–10 days at 10°C instead of the 13–18 days normally required. Kos (1998) has suggested that the cost of upgrading a facility for year-round nitrification with this new process could be as low as 10–15% of the cost for achieving conventional nitrification.

Although kinetic theory provides a basis for bioaugmentation with supplemental nitrifiers developed from high-strength ammonia wastewaters, application requires investigation of several critical concepts. The first of these involves the conditions at which the side-stream (or seed) system must be operated. Because the nitrification process is sensitive to a variety of chemical inhibitors and environmental conditions, especially temperature and pH, the potential for unfavorable conditions for the development of a nitrifying population in a high-strength ammonia wastewater should be evaluated. In addition, the start-up and operating conditions (e.g., feed concentration, SRT) and quantity of nitrifiers that may be produced (i.e., yield) should be evaluated to determine if a sufficient quantity of nitrifiers can actually be developed in the side-stream system to achieve nitrification in the mainstream activated sludge system. Lastly, the ability of the supplemental nitrifiers to rapidly adapt to the mainstream activated sludge system should be evaluated.

A fundamental requirement for successful bioaugmentation is the ability of the separately cultured nitrifiers to rapidly acclimate to new environmental conditions presented by the mixed-culture activated sludge system. The quantity of supplemental nitrifiers must be sufficient to offset nitrifiers washed out of the actual activated sludge system. Therefore, the acclimation period—if present—will affect the quantity of supplemental nitrifiers required.

1.2 Nitrification Kinetics

Ammonium and nitrite oxidation can generally be derived from the growth kinetics of *Nitrosomonas* and *Nitrobacter* (U.S. EPA, 1993). *Nitrosomonas* growth is limited by the concentration of ammonium and *Nitrobacter* growth is limited by the concentration of nitrite as long as inhibitory or toxic conditions do not exist. Nitrite does not usually accumulate in large amounts under steady-state conditions because the maximum growth rate of *Nitrobacter* is considerably higher than the maximum growth rate of *Nitrosomonas*. Therefore, the rate of nitrifier growth is commonly modeled with a Monod-type expression using the conversion of ammonium to nitrite as the rate-limiting step:

$$\mu_N = \mu_{Nmax} \frac{N}{K_N + N} \qquad (1.2\text{-}1)$$

where:
μ_N = specific growth rate of *Nitrosomonas*, d^{-1}
μ_{Nmax} = maximum specific growth rate of *Nitrosomonas*, d^{-1}
K_N = half-saturation coefficient for *Nitrosomonas*, mg/L NH_3-N
N = NH_3-N concentration, mg/L

The rate of ammonium oxidation is controlled by the growth of *Nitrosomonas* and is related to the yield coefficient as follows:

$$q_N = \frac{\mu_N}{Y_N} = q_{Nmax} \frac{N}{K_N + N} \qquad (1.2\text{-}2)$$

where: q_N = ammonium oxidation rate, g NH_3-N oxidized/g volatile suspended solids (VSS) per day
q_{Nmax} = maximum ammonium oxidation rate, g NH_3-N oxidized/g VSS per day
Y_N = yield coefficient, g *Nitrosomonas* grown (VSS)/g NH_3-N removed

As a result of the slow growth rate of nitrifiers compared to heterotrophic microorganisms, a sufficient solids retention time (or mean cell residence time) is essential. The solids retention time is defined as follows:

$$\theta_c = \frac{\text{total mass of biological solids}}{\text{total mass of biological solids leaving the system}} \tag{1.2-3}$$

where: θ_c = solids retention time or mean cell residence time, d

At steady state, the growth rate and solids retention time are related as follows:

$$(\theta_c)^{-1} = \mu_N - b_N = \mu_N' \tag{1.2-4}$$

where: μ_N' = net specific growth rate of nitrifiers, d^{-1}
b_N = endogenous decay coefficient, d^{-1}

For nitrifying organisms, the endogenous decay coefficient is often considered to be negligible ($\mu_N = \mu_N'$).

If the nitrification rate is zero order ($K_N << N$) with respect to ammonium concentration, the ammonium oxidation rate can be expressed as follows:

$$q_N = \frac{\mu_N}{Y_N} = q_{Nmax} \tag{1.2-5}$$

$$q_N = \frac{N_0 - N_1}{X_N t} \tag{1.2-6}$$

where: $N_0 - N_1$ = NH_3-N oxidized, mg/L
X_N = reactor active *Nitrosomonas* as VSS, mg/L
t = hydraulic detention time, d

Typical yield values observed from experimental data for the oxidation of ammonium to nitrite by *Nitrosomonas* were 0.04–0.13 g VSS/g NH_3-N oxidized. For the oxidation of nitrite to nitrate by *Nitrobacter*, typical values were 0.02–0.07 g VSS/g NH_3-N oxidized. Therefore, the total observed yield for complete nitrification was 0.06–0.20 g VSS/g NH_3-N oxidized (U.S. EPA, 1993).

1.3 Selected Literature Related to Nitrification Bioaugmentation

Nitrification kinetics are an essential consideration in the design of an activated sludge process. Kinetic parameters determine the necessary SRT and aeration volume to achieve an effluent goal (Stensel et al., 2002). The SRT is directly related to the maximum specific growth rate. Site-specific assessments of nitrification kinetics are generally necessary because of considerable variability in reported values for the maximum specific growth rate. In addition, nitrifiers are known to be sensitive to environmental conditions such as temperature, pH, and dissolved oxygen (DO). These issues are particularly important when considering nitrification of high-strength ammonia wastewater and nitrifier bioaugmentation into non-nitrifying activated sludge systems.

Kos et al. (2000) demonstrated low-temperature nitrification with a short SRT. Their research indicated that the transport of nitrifying sludge from a warm sidestream process to a colder mainstream process did not pose a problem. Nitrification reaction rates were reduced slightly less than theoretical predictions. Head and Oleszkiewicz (2002) evaluated short SRT nitrification by seeding nitrifying bacteria into cold, non-nitrifying sequencing batch reactors.

The purpose of this work was to determine whether nitrifiers grown on dewatered biosolids centrate at 20°C, 25 °C, and 30°C could induce nitrification in sequencing batch reactors operated at 10°C with an SRT of 4 days. With a specific seed rate and acclimation to 20°C, full nitrification was achieved at hydraulic detention times greater than 43 hr. When seed was discontinued, effluent ammonia concentrations increased. When the hydraulic detention time was reduced to 24 hr partial nitrification was observed. At a hydraulic detention time of 12 hr, no nitrification was observed.

De Silva et al. (2002) studied the effect of nitrifier bioaugmentation in a stressed wastewater treatment system. High-stress conditions were created by increasing the hydraulic loading rate. Experimental and model trends indicated that the maximum effect of a single bioaugmentation was dependent on dose. Performance decreased as bioaugmented nitrifiers were washed out of the system confirming that maintenance doses are necessary. The results suggested a lag before the bioaugmentation procedure became completely effective. Abeysinghe et al. (2002) evaluated the effectiveness of bioaugmentation in nitrifying systems stressed by a washout condition and cold temperature. A short SRT (2 days at 22°C) or very low temperature (4°C with an SRT of 5 days) was used to create high-stress conditions. To achieve low ammonia concentrations, most of the nitrifier population had to be supplied via bioaugmentation (72% of the nitrifier population at the 2-day SRT at 22°C and 95% for the 5-day SRT at 4°C). These large bioaugmentation doses were practically related to poor adaptation to low temperature. De Silva et al. concluded that the process is technically feasible but may not be economically feasible.

Membrane bioreactors have been studied as a means to increase the SRT of activated sludge systems to allow nitrification. Soriano et al. (2003) studied the behavior of conventional activated sludge compared to the membrane bioreactor process under limiting conditions (low SRT and hydraulic residence time). Their research indicated that the membrane prevents washout of nitrifiers. In addition, higher nitrification kinetics and nitrogen removal were observed. Constantine et al. (2001) presented a new process configuration based on the concept of bioaugmentation to achieve nitrification at lower SRTs. The process includes a membrane bioreactor system as the second stage to one train of a multiple-train non-nitrifying activated sludge system. Complete nitrification occurs in the membrane bioreactor system. Biological solids from the membrane bioreactor system are used as the augmentation source for the remainder of the plant. They proposed that this process would require 23% of the tankage for a traditional upgrade.

Inhibitory chemical conditions present an environmental consideration requiring evaluation when nitrifying a high-strength ammonia wastewater. Un-ionized ammonia has been reported to be inhibitory to *Nitrosomonas* at concentrations of 8–124 mg NH_3-N/L and to *Nitrobacter* at concentrations of 0.08–0.82 mg NH_3-N/L (Anthonisen et al., 1976). In addition, un-ionized nitrous acid has been reported to begin to inhibit both *Nitrosomonas* and *Nitrobacter* at concentrations of 0.07–8.83 mg HNO_2-N/L (Anthonisen et.al., 1976).

Noto et.al. (1998) investigated the complete oxidation of high-concentration ammonia (500 mg/L NH_4-N) by retaining incompatible nitrification activities in a three-vessel system. In this system, approximately one-half of the ammonia supplied was oxidized in the first vessel with the remaining one-half oxidized in the second vessel. Nitrite oxidation was not detected in the first two vessels. Nitrite oxidation occurred at high levels in the third vessel suggesting that the processes for oxidizing ammonia and nitrite may be incompatible.

Surmacz-Gorska et al. (1997) studied nitrogen removal from wastewater with a high ammonia concentration via shorter nitrification and denitrification. In this process, ammonium oxidation is promoted while nitrite oxidation is intentionally inhibited. Denitrification begins with nitrite as the initial substrate, not nitrate. Their research indicated that *Nitrobacter* are sensitive to free ammonium, the presence of which depends on the total ammonia concentration and pH. Nitrite accumulation occurred at concentrations of free ammonium at 1–6 mg NH_3/L.

Mossakowska et al. (1997) investigated nitrification rates in the treatment of supernatant from dewatering of digested sludge. Their studies of the nitrification process found that nitrite accumulated until ammonium was depleted. After all ammonium was oxidized, all nitrite could thereafter be completely oxidized to nitrate. It was shown that nitrite accumulation was primarily caused by the initial concentration of ammonium and that oxygen concentration was secondary.

Previous research provides both a basis for nitrifier bioaugmentation in non-nitrifying activated sludge system as well as considerations that require evaluation in designing and operating a full-scale system.

1.4 Research Objective

The research team investigated the potential to culture nitrifying microorganisms in a high-strength ammonia substrate using high-concentration start-up conditions and batch-feed operation, and the ability of these cultures to adapt or acclimate to new conditions when used to bioaugment a non-nitrifying activated sludge process operated at similar environmental conditions (e.g., temperature, pH) to achieve nitrification.

This laboratory bench-scale investigation was conducted in two phases. In the first phase, cultures of nitrifying microorganisms were developed in a high-strength ammonia synthetic wastewater and in dewatered biosolids supernatant. In the second phase, these cultures were used in bench-scale bioaugmentation analyses to evaluate acclimation. Nitrifiers obtained from a nitrifying trickling filter facility were also used in these analyses for comparison. Bioaugmentation analyses were conducted in the mixed liquor suspended solids (MLSS) and final clarifier effluent from a non-nitrifying activated sludge system.

This investigation is a preliminary step in evaluating bioaugmentation in an activated sludge process with supplemental nitrifies intended to provide a technical basis for potential continuous flow pilot-scale studies or full-scale systems.

CHAPTER 2.0

EXPERIMENTAL METHODS

2.1 Waste Characteristics

For subsequent bioaugmentation, seed nitrifying cultures were developed from commercially available nitrifiers grown in a high-strength ammonia synthetic wastewater [~1000 mg/L ammonia-nitrogen (NH_3-N)]. Before bioaugmentation in activated sludge, experimental procedures were verified in a low-strength (~25 mg/L NH_3-N) synthetic wastewater obtained by diluting the high-strength wastewater. To ensure that limiting factors for nitrification were not introduced, a brief literature search was conducted on the appropriate characteristics for the synthetic wastewater. The characteristics of the high-strength synthetic wastewater used are shown in Table 2-1.

Table 2-1. Synthetic Wastewater Characteristics.

Constituent	Concentration (mg/L)
$MgSO_4$-$7H_2O$	20,000
$FeSO_4$-$7H_2O$	2,500
$ZnSO_4$-$7H_2O$	200
$MnSO_4$-H_2O	30
$CaCl_2$-$2H_2O$	2,000
Na_2HPO_4	2,481
KH_2PO_4	1,215
$NaHCO_3$	2,453
$(NH_4)_2SO_4$	4,717

Seed nitrifying populations for subsequent bioaugmentation were also developed from microorganisms indigenous to dewatered biosolids supernatant. This supernatant typically contains high concentrations of ammonia. Bioaugmentation procedures were verified in diluted supernatant. An analysis of pertinent constituents in dewatered biosolids supernatant obtained from the City of Moorhead (Minn.) Wastewater Treatment Facility was conducted for comparison to the synthetic waste. Typical characteristics of the undiluted dewatered biosolids supernatant are shown in Table 2-2.

Table 2-2. Typical Dewatered Biosolids Supernatant Characteristics.

Parameter	Concentration (mg/L)
Mg	12.9
Fe	3.04
Zn	0.292
Mn	0.087
Ca	55.1

2.2 Sources of Microorganisms

Two sources for cultures of nitrifying microorganisms were used. A vendor for commercially available nitrifiers was identified and samples were obtained. Nitrifiers from a nitrifying trickling filter facility were obtained locally from the City of Fargo (N.D.) Wastewater Treatment Facility. Non-nitrifying activated sludge samples were obtained from the City of Moorhead (Minn.) Wastewater Treatment Facility. This facility uses a high-purity oxygen activated sludge system that is typically operated at an SRT of 3 days.

2.3 Bench-scale Bioreactors

A schematic of the bench-scale system is shown in Figure 2-1. The clear Plexiglas bench-scale bioreactors had a diameter of 19.0 cm and a height of 27.0 cm. The reactors were operated at a liquid depth of 17.5 cm. The total reactor volume was 7.7 L with a working volume of 5.0 L. Two air diffusers were located 180° apart at a height of 2.5 cm. Air flow to the reactors first passed through a humidifier. Magnetic mixers were used in addition to aeration to ensure complete mixing. The DO concentration was maintained at 7–8 mg/L. Samples of synthetic wastewater and/or biosolids supernatant were fed through the top of the reactor. Samples of effluent and biomass were removed through an access port located at a height of 7.5 cm. During operation, the bioreactors are placed in an incubator maintained at approximately 20°C.

2.4 Development of Cultures of Nitrifiers for Bioaugmentation

For subsequent bioaugmentation, nitrifying cultures were developed from commercially available nitrifiers grown in the high-strength synthetic wastewater and from microorganisms indigenous to dewatered biosolids supernatant.

Three bioreactors were started with 5.0 L of high-strength synthetic wastewater seeded with a predetermined quantity of commercially available nitrifiers based on supplier recommendations. A fourth bioreactor was started with 5.0 L of dewatered biosolids supernatant. Feeding and wasting procedures were conducted at approximately the same time, once per day. The first three reactors were fed with high-strength ammonia synthetic wastewater that had been maintained at approximately the same temperature and pH as the reactor. The fourth reactor was fed with dewatered biosolids supernatant preserved at 4°C. The alkalinity of the supernatant was adjusted to ensure that adequate alkalinity was present for nitrification. Process start-up was conducted by feeding high-strength ammonia wastewater directly (versus start-up using lower concentrations and a gradual increase to higher concentrations).

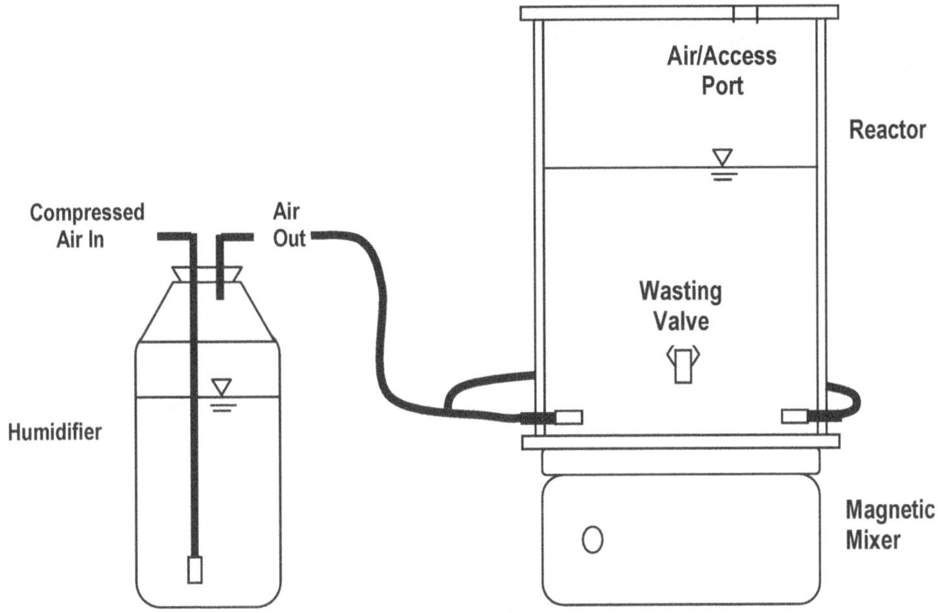

Figure 2-1. Bench-scale System.

Biomass removal was conducted during aeration and mixing to maintain an SRT equivalent to the hydraulic retention time. The three reactors using synthetic wastewater were operated at three different SRT values: 5 days, 10 days, and 20 days. The reactor using dewatered biosolids supernatant was operated at an SRT of 20 days.

Routine influent monitoring included analyses for soluble chemical oxygen demand (COD), NH_3-N, total suspended solids (TSS), VSS, pH, and alkalinity. Routine effluent monitoring included analyses for COD (when interference was not present), ammonia-nitrogen, nitrite-nitrogen (NO_2-N), nitrate-nitrogen (NO_3-N), TSS, VSS, pH, alkalinity, and DO. Parameters were measured at a frequency ranging from once per day to once per week, as indicated in Table 2-3.

2.5 Bioaugmentation Analyses

Bioaugmentation procedures were verified by evaluating acclimation and ammonia removal in low-strength synthetic wastewater or diluted dewatered biosolids supernatant supplemented with separately cultured nitrifiers developed from each respective substrate. Evaluation of acclimation and ammonia removal in a clarified effluent from a non-nitrifying activated sludge (i.e., final clarifier effluent) and in non-nitrifying activated sludge supplemented with separately cultured nitrifiers or trickling filter nitrifiers (via direct bioaugmentation) were conducted after bioaugmentation procedures were verified.

Table 2-3. Routine Monitoring for Seed Bioreactors.

Sample	Parameter	Frequency
Influent	Volume	1/day
	COD	1/week
	NH_3-N	1/week
	TSS/VSS	1/week
	pH	1/week
	Alkalinity	1/week
Effluent and wasted biomass	Soluble COD	1/week
	NH_3-N	3/week
	NO_2-N	3/week
	NO_3-N	3/week
	TSS/VSS	2/week
	pH	5/week
	Alkalinity	3/week
	DO	3/week

2.5.1 Verification of Bioaugmentation Procedures in Low-strength Waste

The research team used U.S. EPA bioreactor systems identical to that shown in Figure 2-1 for bioaugmentation analyses. The bioreactors and respirometer (for concurrent oxygen uptake analyses) were maintained in an incubator at a constant temperature of approximately 20°C during analysis. Oxygen uptake analyses using the respirometer were conducted on selected test runs.

To conduct the verification evaluation, a predetermined volume of low-strength synthetic wastewater or diluted dewatered biosolids supernatant (depending on the source of supplemental nitrifiers) was added to the bioreactor and respirometer chambers. The ammonia-nitrogen concentration of the diluted substrate was 17–27 mg/L. A known quantity of biomass (volume and VSS basis) was obtained from one of the seed bioreactors used to separately culture nitrifiers or from trickling filter biomass. The total combined volume (substrate and biomass) used in the bioreactor was 4.0–6.0 L. For the respirometer, the total combined volume was based on the anticipated oxygen demand.

To evaluate acclimation and ammonia removal, aliquots were removed and analyzed for temperature, pH, DO, ammonia-nitrogen, nitrite-nitrogen, and nitrate-nitrogen. TSS and VSS analyses were also conducted at varying time intervals. Samples were collected approximately hourly depending on the observed ammonia removal rate. Cumulative oxygen uptake was recorded automatically every 0.96 hr.

2.5.2 Bioaugmentation in Clarified Effluent from a Non-nitrifying Activated Sludge

The procedure outlined above was repeated using clarified effluent from a non-nitrifying activated sludge (i.e., final clarifier effluent).

2.5.3 Bioaugmentation in Non-nitrifying Activated Sludge

For bioaugmentation analyses in non-nitrifying activated sludge, a sample of the MLSS was collected from the influent end of an activated sludge aeration basin at the Moorhead Wastewater Treatment Facility (immediately after the addition of return activated sludge). The return activated sludge at this facility typically has a low pH because of saturation with carbon dioxide. Therefore, the sample was briefly aerated to strip carbon dioxide and increase the pH to near 7.

The research team used two bioreactors and four respirometer chambers (each in duplicate). A predetermined volume of MLSS was mixed with a known quantity of biomass (volume and VSS basis) obtained from one of the seed bioreactors used to separately culture nitrifiers or from trickling filter biomass. A second bioreactor and pair of respirometer chambers were used as a control, and therefore, no cultured biomass was added to the activated sludge. The purpose of the control was to establish a baseline that incorporated the potential for ammonia-nitrogen production (via ammonification of organic nitrogen) and removal (via uptake in the biomass) in the activated sludge. A total combined volume (substrate and biomass) of 4.0–6.0 L was used in the bioreactor. For the respirometer, the total combined volume was based on the anticipated oxygen demand. Oxygen uptake analyses were conducted on selected test runs. Control bioreactors were not used in analyses with trickling filter biomass.

To evaluate acclimation and ammonia removal, samples were removed and analyzed for temperature, pH, DO, ammonia-nitrogen, nitrite-nitrogen, and nitrate-nitrogen. TSS and VSS analyses were also conducted at varying time intervals. Samples were collected approximately hourly depending on the observed ammonia removal rate. Cumulative oxygen uptake was recorded automatically every 0.96 hr.

2.5.4 Bioaugmentation Data Evaluation

Two criteria were established to investigate the ability of supplemental nitrifiers to adapt to new conditions when used for bioaugmentation as well as the possibility of an acclimation period:

1. Ammonia removal, nitrite production, and nitrate production were observed for several hours after bioaugmentation. If the initial ammonia removal, nitrite production, or nitrate production rate was significantly slower than that observed after the initial period, as determined by the removal and/or production rate, this was interpreted as an acclimation period. No significant change in the removal and/or production rates was interpreted as indicating that acclimation was not required.
2. The ammonium oxidation rate (q_N) was computed for the bioaugmentation analysis and compared to that from the seed bioreactor from which the supplemental nitrifiers were obtained. A q_N value significantly lower in the bioaugmentation analysis was interpreted as indicating the need for an acclimation period. A q_N value from the bioaugmentation analysis approximately equal to or greater than that observed in the seed bioreactor was interpreted as indicating that no acclimation was needed.

2.6 Analytical Methods

Respirometric analyses were conducted using a BODTrak respirometer (Hach Company, Loveland, Colo.). All other parameters were measured in accordance with *Standard Methods* (APHA, 1998).

CHAPTER 3.0

DEVELOPMENT OF CULTURES OF NITRIFIERS FOR BIOAUGMENTATION

3.1 High-strength Ammonia Synthetic Waste

3.1.1 5-day SRT

The bioreactor using high-strength ammonia synthetic waste at a 5-day SRT operated for 76 days. The temperature and DO in the bioreactor were maintained at 20.0°C ± 2.0°C and 8.3 mg/L ± 1.1 mg/L, respectively. Results from routine monitoring are shown in Figures 3-1 through 3-4.

Although pH did not appear to significantly inhibit nitrification, pH adjustment was initiated on day 33 to ensure that a favorable pH was maintained. In Figure 3-1, the effluent pH represents the pH of the effluent withdrawn before daily influent addition. The adjusted bioreactor pH represents the pH in the bioreactor after the addition of influent and alkalinity. The difference represents the change in pH that occurred over the course of daily operation.

The lowest effluent ammonia concentrations, shown in Figure 3-2, were observed from days 35–65. Although the nitrite concentrations increased in relative proportion to the decrease in the ammonia concentration, complete conversion to nitrate was not observed.

Un-ionized ammonia has been reported to be inhibitory to *Nitrosomonas* at concentrations of 8–124 mg NH_3-N/L and to *Nitrobacter* at concentrations of 0.08–0.82 mg NH_3-N/L (Anthonisen et al., 1976). In addition, un-ionized nitrous acid has been reported to begin to inhibit both *Nitrosomonas* and *Nitrobacter* at concentrations of 0.07–8.83 mg HNO_2-N/L (Anthonisen et al., 1976). Concentrations of un-ionized ammonia-nitrogen in the bioreactor and in the effluent are shown in Figure 3-3. The bioreactor un-ionized ammonia-nitrogen concentration represents the concentration immediately after influent addition, which decreased to the effluent concentration over the course of daily operation.

As indicated in Figure 3-3, effluent un-ionized ammonia concentrations dropped below inhibitory concentrations for *Nitrosomonas* on approximately day 35 and remained at that level until day 65. During these 30-days, ammonia conversion remained relatively constant. After day 65, un-ionized ammonia concentrations increased coincident with an increase in effluent pH and effluent ammonia, and subsequently, a decrease in effluent nitrite was observed. For the purpose of evaluating kinetic parameters, days 35–65 were considered to be quasi-steady-state operation. Un-ionized ammonia concentrations remained well above the referenced inhibitory concentrations for *Nitrobacter*, and therefore, conversion of nitrite to nitrate was not observed.

As shown in Figure 3-4, effluent alkalinity decreased from greater than 2000 mg/L to slightly less than 1500 mg/L after day 30. The alkalinity then remained relatively constant through day 65. Sufficient alkalinity was maintained for nitrification. Using the effluent VSS concentration as an approximate measure of biological solids, the 5-day SRT did not appear to support sufficient biological growth to achieve a greater degree of nitrification under the operating conditions evaluated. Concentrations of un-ionized ammonia inhibitory to *Nitrobacter* were consistently observed. Therefore, operation of this bioreactor was discontinued on day 76.

For the conditions evaluated, it appears that a sequencing batch reactor system with daily feed operated at a 5-day SRT may not be a practical operational regime to develop the quantity of nitrifiers necessary for supplemental bioaugmentation into an activated sludge process. However, other operating strategies with a 5-day SRT (not evaluated in this study) may be feasible and/or warrant further study. Process start-up using a lower influent ammonia concentration with a gradual increase to a high-strength influent ammonia concentration and/or a continuous feed (versus batch feed) may alleviate inhibitory conditions and allow steady-state operation to be achieved. At a minimum, the design configuration and operational strategy for a system to operate at a 5-day SRT must consider the potential for inhibition caused by un-ionized ammonia. In addition, the degree to which nitrification is complete is also a consideration. Incomplete nitrification (e.g., 20–30% ammonia oxidation as indicated in Figure 3-2) will result in significant ammonia carryover to the activated sludge system, thereby increasing the loading to the system.

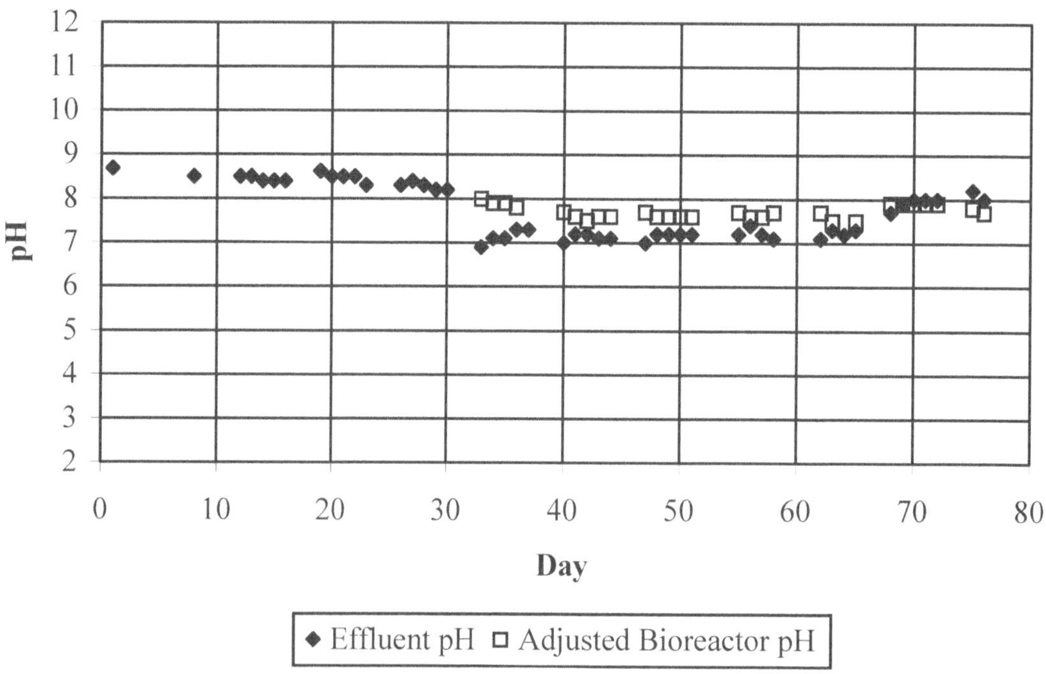

Figure 3-1. Effluent and Adjusted Bioreactor pH; 5-day SRT.

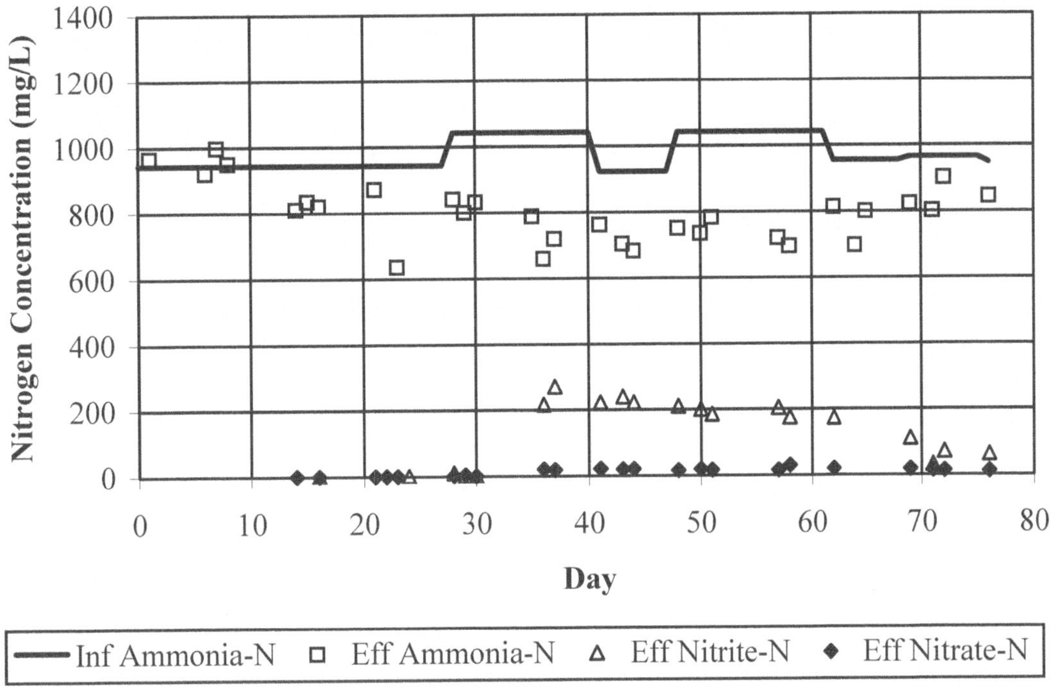

Figure 3-2. Nitrogen Concentrations; 5-day SRT (Inf, influent; Eff, effluent).

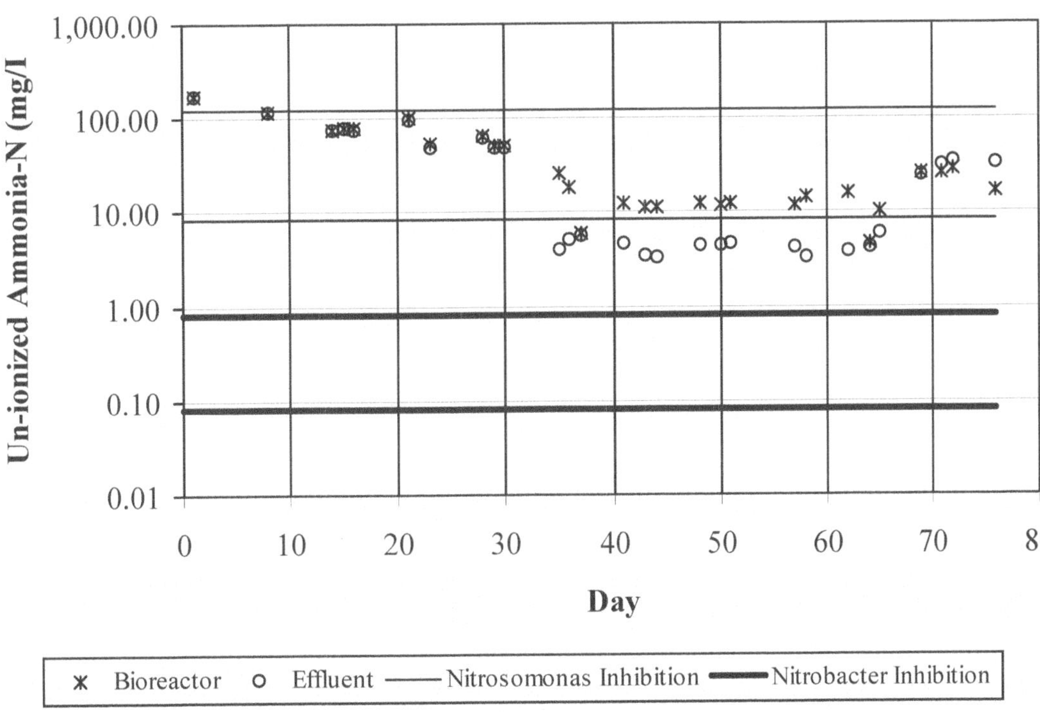

Figure 3-3. Un-ionized Ammonia-nitrogen Concentrations; 5-day SRT.

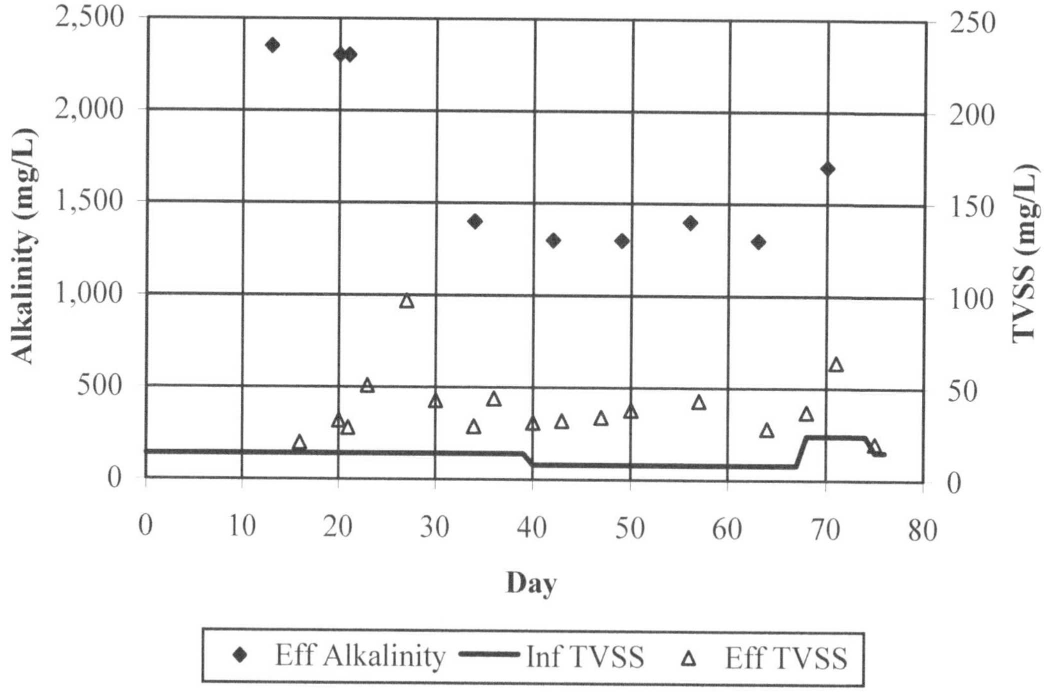

Figure 3-4. Alkalinity and Total Volatile Suspended Solids (TVSS); 5-day SRT (Inf, influent; Eff, effluent).

3.1.2 10-day SRT

The 10-day SRT bioreactor operated for 252 days. The temperature and DO in the bioreactor were maintained at 20.2°C ± 1.9°C and 8.0 mg/L ± 3.2 mg/L, respectively. Results from routine monitoring are shown in Figures 3-5 through 3-9.

As indicated in Figure 3-5, without adjustment, the effluent pH dropped to less than 7 by day 26. To ensure that a pH favorable for nitrification was maintained, pH adjustment was initiated on day 26. The lowest effluent pH was observed between day 26 and day 68, coincident with a rapid decrease in the effluent ammonia concentration (Figure 3-6). After approximately day 68 the adjusted bioreactor pH was maintained nearly constant; the effluent pH varied from slightly less than 7 to approximately 8.5.

Effluent ammonia-nitrogen concentrations steadily decreased from approximately 1000 mg/L to less than 200 mg/L between days 0 and 60. On day 28, the effluent un-ionized ammonia concentration (indicated in Figure 3-7) dropped to a level that was no longer inhibitory to nitrite formation (by *Nitrosomonas*), and therefore, the nitrite concentration began to increase as shown in Figure 3-6. As nitrite was produced un-ionized nitrous acid was also being formed. In Figure 3-8, the bioreactor un-ionized nitrous acid concentration represents the concentration immediately after influent addition, which decreased to the effluent concentration over the course of daily operation. The effluent un-ionized nitrous acid concentration subsequently reached a level inhibitory to *Nitrosomonas* and the continued production of nitrite. As a result, the effluent ammonia concentration increased after day 60. In addition, after day 60 the effluent nitrite concentration began to slowly decrease, thereby reducing the concentration of un-ionized nitrous acid. After approximately day 110 it appears that the un-ionized nitrous acid concentration had decreased to a level that was no longer inhibitory to either *Nitrosomonas* or *Nitrobacter*. The result was a decrease in effluent ammonia-nitrogen to a relatively stable concentration of approximately 200 mg/L by day 150. Between days 110 and 150, effluent nitrite concentrations continued to decrease coincident with significant increases in the effluent nitrate concentration. The concentration of un-ionized nitrous acid was low enough that *Nitrobacter* activity was not inhibited during this period. Although concentrations of un-ionized ammonia remained at levels potentially inhibitory to *Nitrobacter*, the population appears to have been sufficiently well established or acclimated to overcome the inhibitory effect. After day 169 it appears that even slight increases in the effluent un-ionized ammonia concentration may have upset the delicate balance and become inhibitory to the continued production of nitrate. As a result, an increase in effluent nitrite accompanied by a decrease in effluent nitrate was observed.

The observed cycling between ammonia removal, nitrite production and removal, and nitrate production indicates the sensitive balance between *Nitrosomonas* and *Nitrobacter* in relation to un-ionized ammonia, un-ionized nitrous acid, and pH associated with the complete conversion of high-strength ammonia to nitrate. Changes in conditions in the bioreactor related to these constituents directly affected the predominant biological activity taking place. When conditions near inhibitory levels were experienced, it appears that well-established or acclimated populations could continue to function beyond inhibitory levels to a certain degree. Similarly, when conditions changed from inhibitory to noninhibitory, a lag was experienced during which the new population of microorganisms became established.

As indicated in Figure 3-9, effluent alkalinity concentrations remained greater than 800 mg/L, which indicated that sufficient alkalinity was maintained to support complete nitrification. The production of biological solids, as measured by effluent VSS concentrations, was slightly greater than that observed for the bioreactor using high-strength synthetic waste operating at a 5-day SRT. On the basis of relatively constant ammonia removal from days 146–252, this period was considered to be quasi-steady-state operation for the purpose of evaluating kinetic parameters. This was concluded although the steady-state conversion of nitrite to nitrate was not consistently observed.

For the conditions evaluated, it appears that a sequencing batch reactor system with daily feed operated at a 10-day SRT may not be a practical operational regime to develop the quantity of nitrifiers necessary for supplemental bioaugmentation into an activated sludge process. However, other operating strategies with a 10-day SRT (not evaluated in this study) may be feasible and/or warrant further study. Process start-up using a lower influent ammonia concentration with a gradual increase to a high-strength influent ammonia concentration and/or a continuous feed (versus batch feed) may alleviate inhibitory conditions and allow steady-state operation to be achieved. At a minimum, the design configuration and operational strategy for a system to operate at a 10-day SRT must consider the potential for inhibition caused by un-ionized ammonia and un-ionized nitrous acid. Incomplete nitrification (e.g., ~80% as indicated in Figure 3-6) will result in ammonia carry-over to the activated sludge system thereby increasing the loading to the system.

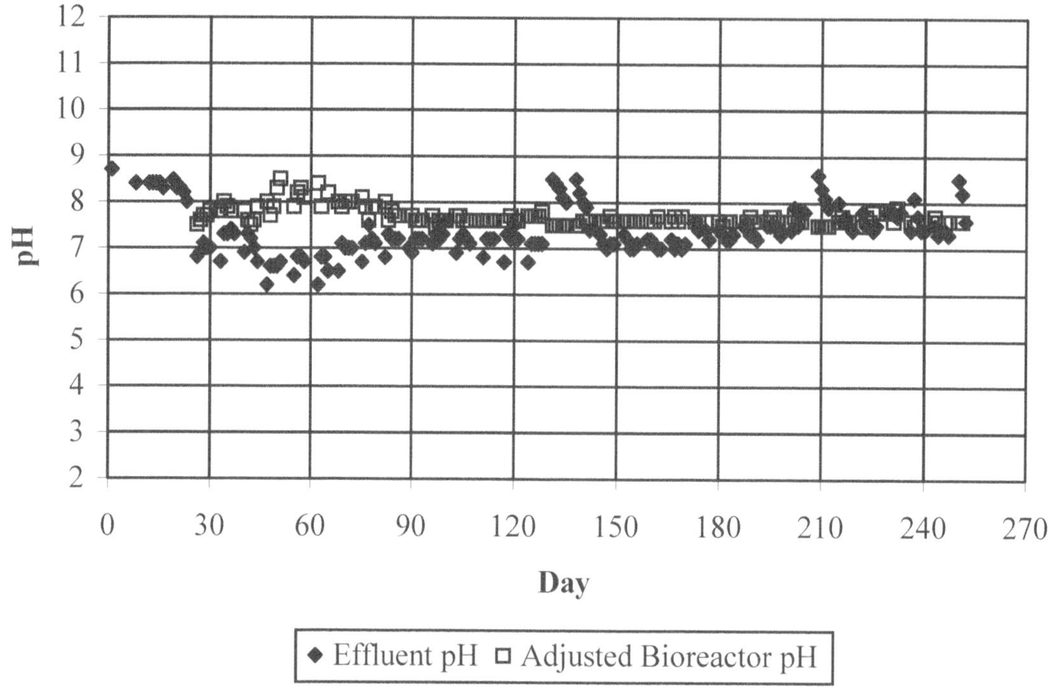

Figure 3-5. Effluent and Adjusted Bioreactor pH; 10-day SRT.

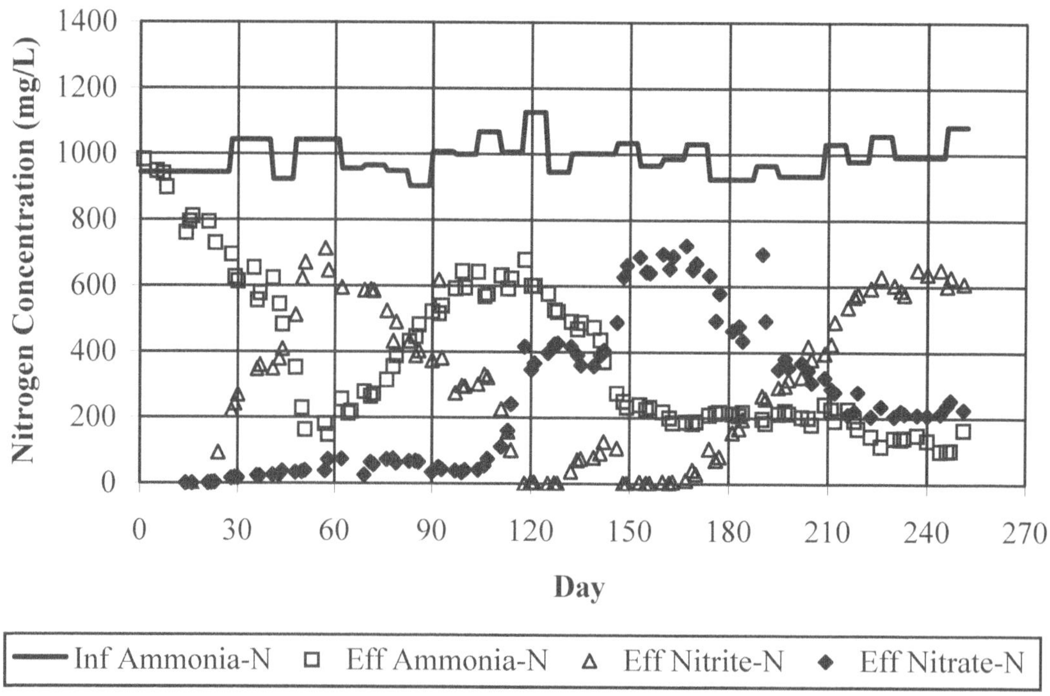

Figure 3-6. Nitrogen Concentrations; 10-day SRT (Inf, influent; Eff, effluent).

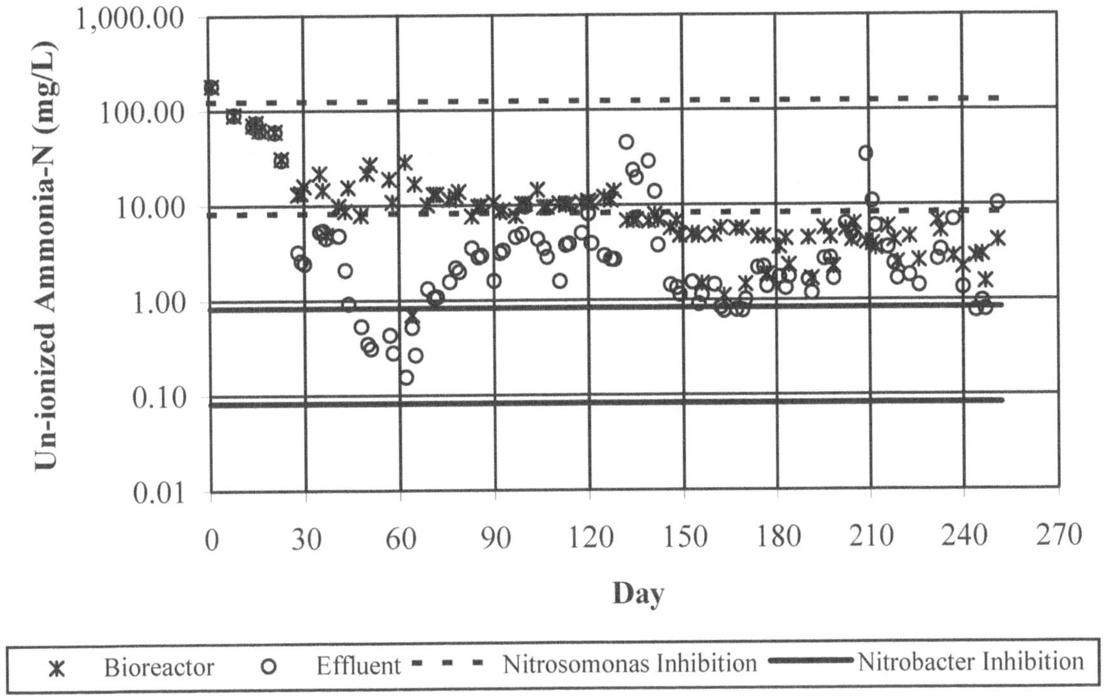

Figure 3-7. Un-ionized Ammonia-nitrogen Concentrations; 10-day SRT.

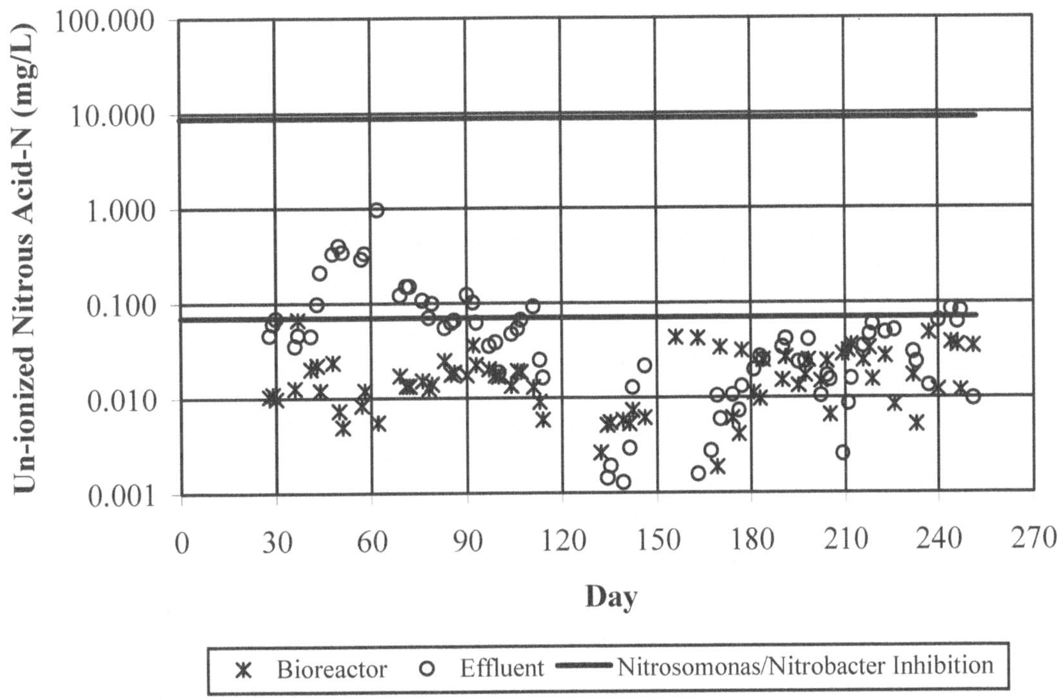

Figure 3-8. Un-ionized Nitrous Acid Concentrations; 10-day SRT.

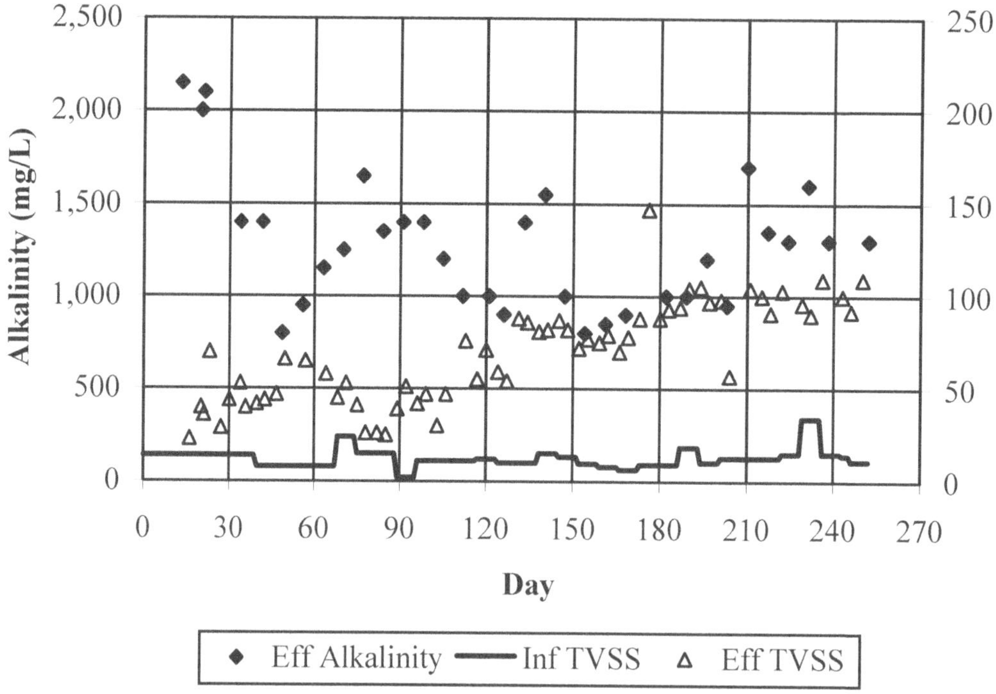

Figure 3-9. Alkalinity and TVSS; 10-day SRT (Inf, influent; Eff, effluent).

3.1.3 20-day SRT

The 20-day SRT bioreactor operated for 260 days. The temperature and DO in the bioreactor were maintained at 20.2°C ± 1.8°C and 8.0 mg/L ± 1.7 mg/L, respectively. Results from routine monitoring are shown in Figures 3-10 through 3-14. Although a 20-day SRT may be longer than the minimum required for complete nitrification, this scenario was used to ensure that adequate biomass for further bioaugmentation studies would be available.

As indicated in Figure 3-10, the effluent pH dropped to less than 7 by day 19. To ensure that a pH favorable for nitrification was maintained, pH adjustment was initiated on day 26. The lowest effluent pH was observed between day 19 and day 41 coincident with a rapid decrease in the effluent ammonia concentration (Figure 3-11). The adjusted bioreactor pH and effluent pH remained nearly constant after approximately day 50.

Effluent ammonia-nitrogen concentrations rapidly decreased from approximately 1000 mg/L to less than 100 mg/L by day 43. After day 16, the effluent un-ionized ammonia concentration (indicated in Figure 3-12) decreased to a level that was no longer inhibitory to *Nitrosomonas.* Therefore, effluent nitrite-nitrogen increased to approximately 800 mg/L by day 50, as indicated in Figure 3-11. As nitrite was produced, the substrate became self-inhibitory as indicated by the effluent un-ionized nitrous acid concentration increasing as shown in Figure 3-13. These inhibitory conditions resulted in a slow increase in effluent ammonia from day 50 through day 114. After day 50, effluent nitrite concentrations began to slowly decrease resulting in effluent un-ionized nitrous acid concentrations decreasing below inhibitory levels by day 97. Despite un-ionized nitrous acid concentrations potentially inhibitory to both *Nitrosomonas* and *Nitrobacter*, continued ammonia removal and nitrite removal was observed between day 50 and day 114. It would appear that the populations of both microorganisms were well established or acclimated so that some degree of activity could continue despite the inhibitory concentrations. After day 97, the concentrations of both un-ionized ammonia and un-ionized nitrous acid began to decrease to a point where the complete conversion of ammonia to nitrate was observed. By day 104 essentially all nitrite was being converted to nitrate. After day 163 effluent ammonia-nitrogen concentrations remained less than 50 mg/L.

For the purpose of evaluating kinetic parameters, days 163–260 were considered to be steady-state operation as indicated by the consistently low effluent ammonia concentrations and near complete conversion to nitrate with essentially no nitrite accumulation. The 20-day SRT produced a steady-state condition suggesting a stable operation capable of providing a viable treatment regime for the near complete conversion of a high-strength ammonia waste to nitrate from which a culture of nitrifiers could be developed for supplemental bioaugmentation in an activated sludge process. However, the data do not indicate that a 20-day SRT is the minimum required. Systems operated at lower SRTs, under specific operational conditions (as previously discussed), may also satisfy this objective.

Effluent alkalinity concentrations decreased to less than 500 mg/L by day 19 coincident with the rapid pH drop. Effluent alkalinity remained greater than 800 mg/L after day 30 indicating that sufficient alkalinity was maintained to support nitrification. The production of biological solids, as measured by effluent VSS concentrations, was slightly greater than that observed for the bioreactors using high-strength synthetic waste operating at the 5-day and 10-day SRT.

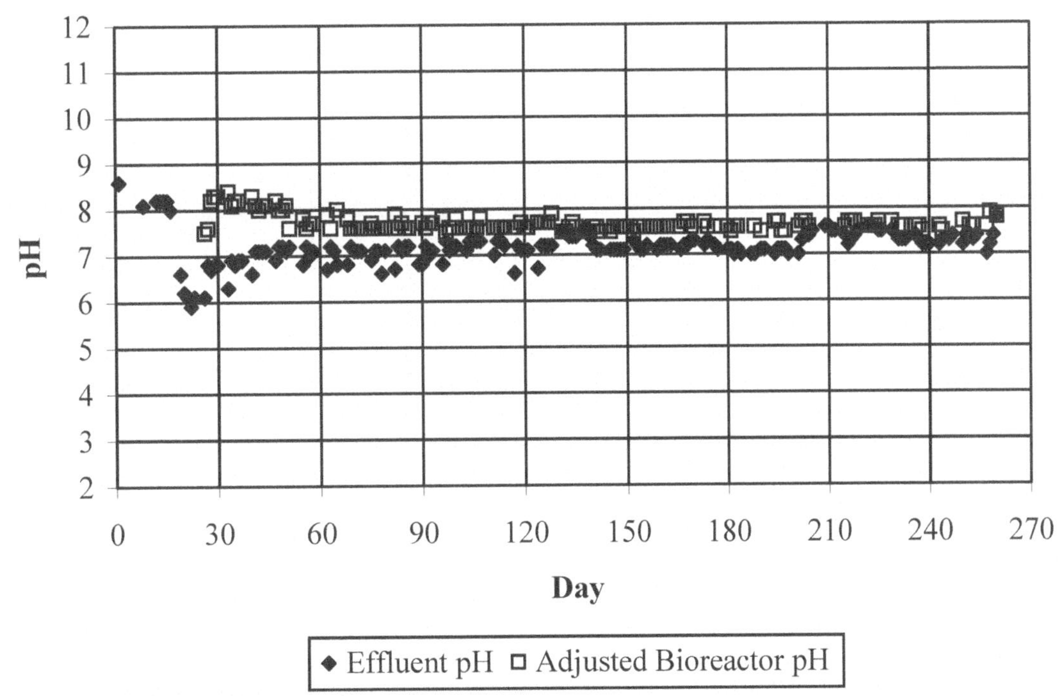

Figure 3-10. Effluent and Adjusted Bioreactor pH; 20-day SRT.

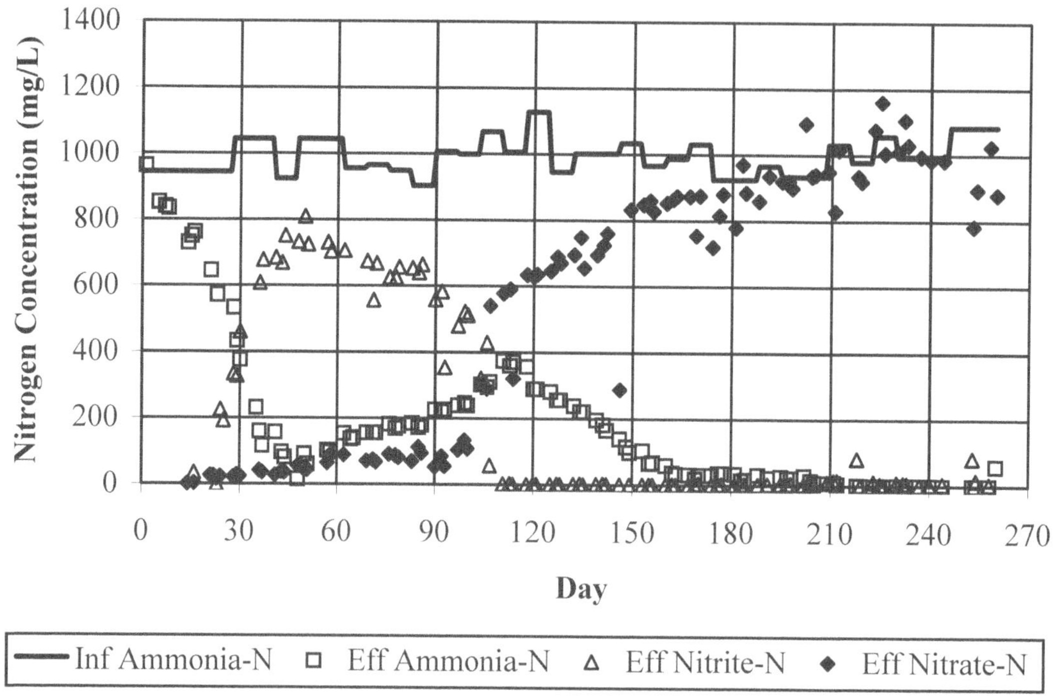

Figure 3-11. Nitrogen Concentrations; 20-day SRT (Inf, influent; Eff, effluent).

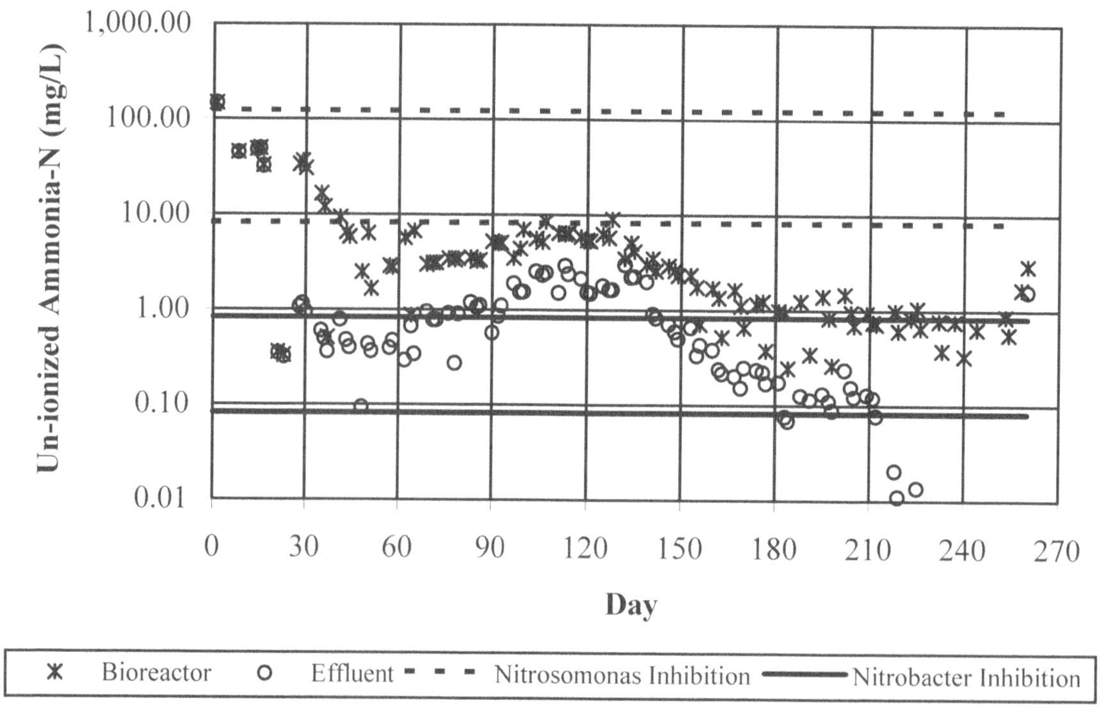

Figure 3-12. Un-ionized Ammonia-nitrogen Concentrations; 20-day SRT.

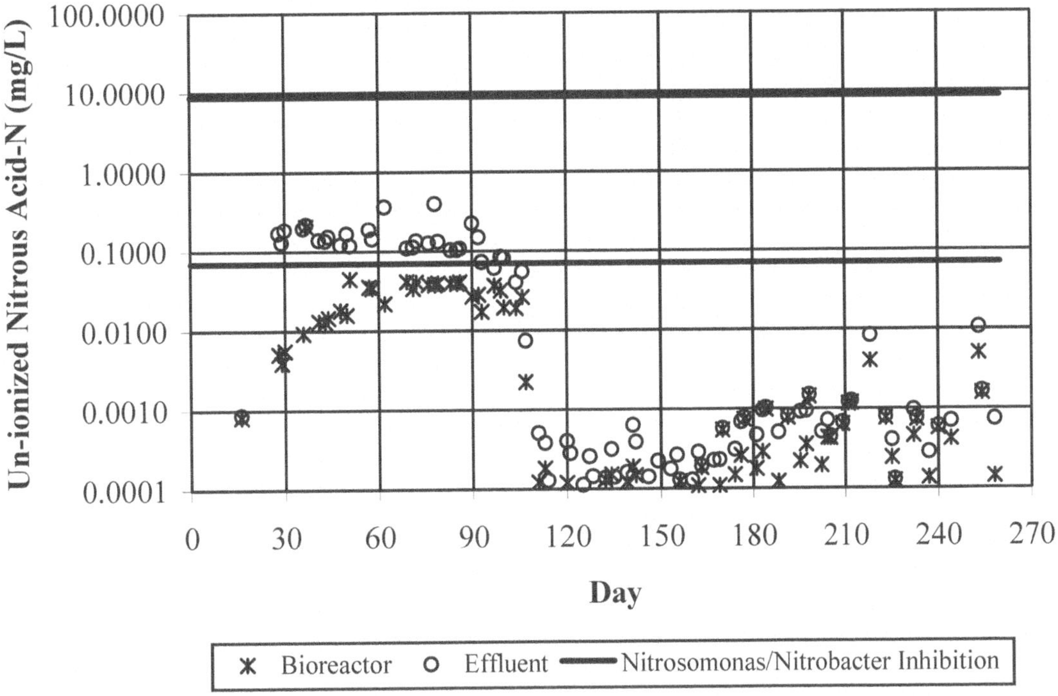

Figure 3-13. Un-ionized Nitrous Acid Concentrations; 20-day SRT.

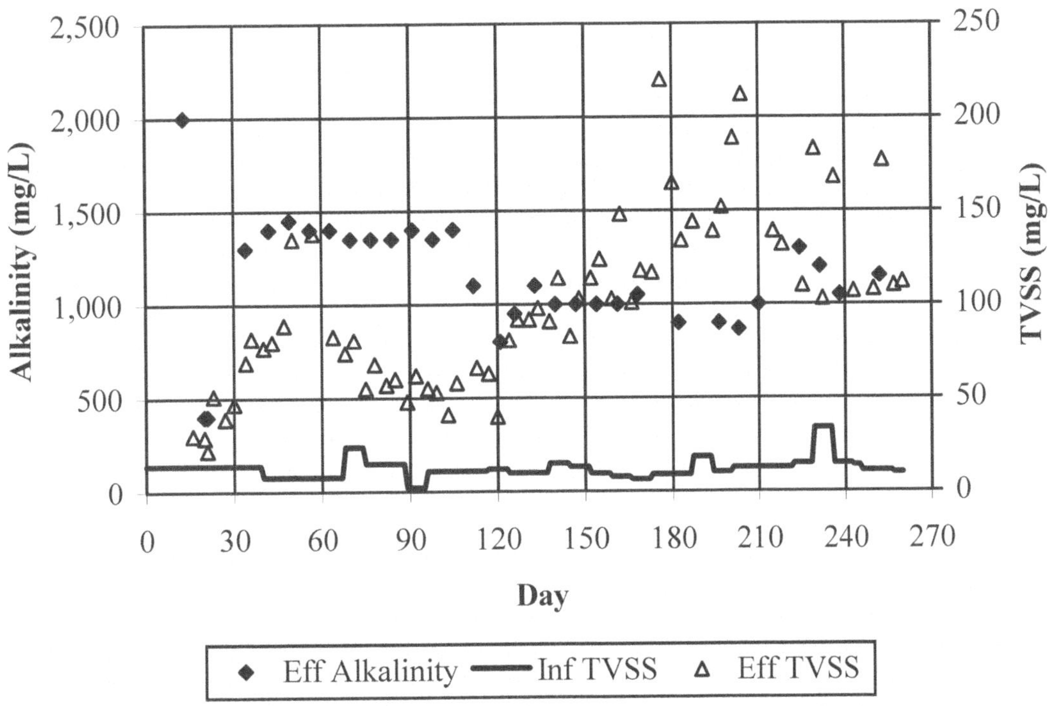

Figure 3-14. Alkalinity and TVSS; 20-day SRT (Inf, influent; Eff, effluent).

3.2 Dewatered Biosolids Supernatant at 20-day SRT

The bioreactor using dewatered biosolids supernatant at a 20-day SRT operated for 250 days. The temperature and DO in the bioreactor were maintained at 20.2°C ± 2.6°C and 8.1 mg/L ± 1.8 mg/L, respectively. Results from routine monitoring are shown in Figures 3-15 through 3-19. Although the 20-day SRT may be longer than the minimum required for complete nitrification, this scenario was used to ensure that adequate biomass for further bioaugmentation studies would be available.

As indicated in Figure 3-15, the effluent pH remained above 9 until approximately day 40. During this period of operation, the effluent ammonia-nitrogen concentration rapidly dropped to near 200 mg/L by day 14 and to less than 40 mg/L by day 41 (Figure 3-16). Conversion to nitrite was observed beginning on day 41. At the high initial effluent pH, it is probable than the ammonia removal was, at least partially, the result of air stripping the gas phase fraction of the total ammonia concentration. Beginning on day 35 pH adjustment was initiated to lower the bioreactor pH to minimize the release of gas phase ammonia. After this initial adjustment the effluent pH dropped to less than 7 by day 50. Adjustment beyond day 50 was conducted to maintain a minimum pH for nitrification. Effluent pH was consistently between 6 and 7 on days 48–106. After day 106, the effluent pH increased to between 8 and 9. During this time, the adjusted bioreactor pH was maintained between 7 and 8.

Effluent ammonia-nitrogen concentrations rapidly decreased as previously described (Figure 3-16). Beginning on day 41 and continuing through approximately day 110, a proportional increase in nitrite was observed with effluent nitrite-nitrogen concentrations increasing to 800–1000 mg/L. As the effluent pH decreased after day 41, the concentration of un-ionized ammonia decreased below inhibitory levels for both *Nitrosomonas* and *Nitrobacter* as shown in Figure 3-17, and remained at that level through day 250. As the nitrite concentrations increased after day 41, un-ionized nitrous acid concentrations also increased as shown in Figure 3-18. However, the concentrations did not appear to be inhibitory to *Nitrosomonas* as increases in effluent ammonia were not observed. Un-ionized nitrous acid concentrations may have been slightly inhibitory to *Nitrobacter* as indicated by the nitrite accumulation. After day 118, near complete conversion of ammonia to nitrate was quickly established and no inhibitory conditions were observed. This operation remained stable although the influent ammonia-nitrogen concentration increased from 1200 to 1700 mg/L. The steady-state condition is reflected in Figure 3-16 with effluent ammonia and nitrite concentrations remaining very low and effluent nitrate concentrations reflecting the influent ammonia concentration.

For the purpose of evaluating kinetic parameters, days 118–250 were considered to be a steady-state operation as indicated by the consistently low-effluent ammonia concentrations and the near-complete conversion to nitrate with essentially no nitrite accumulation. The 20-day SRT produced a steady-state condition. This condition suggests a stable operation capable of complete conversion of a high-strength ammonia waste to nitrate from which a culture of nitrifiers could be developed for supplemental bioaugmentation in an activated sludge process. However, the data do not indicate that a 20-day SRT is the minimum required. Under specific operational conditions, systems operated at lower SRTs may also satisfy this objective.

Effluent alkalinity concentrations remained greater than 150 mg/L indicating that sufficient alkalinity was maintained to support nitrification. The average effluent VSS concentration in the bioreactor during days 118–250 was 670 mg/L. However, the presence of

VSS not associated with biological solids likely renders this parameter ineffective as a measure of bioreactor performance.

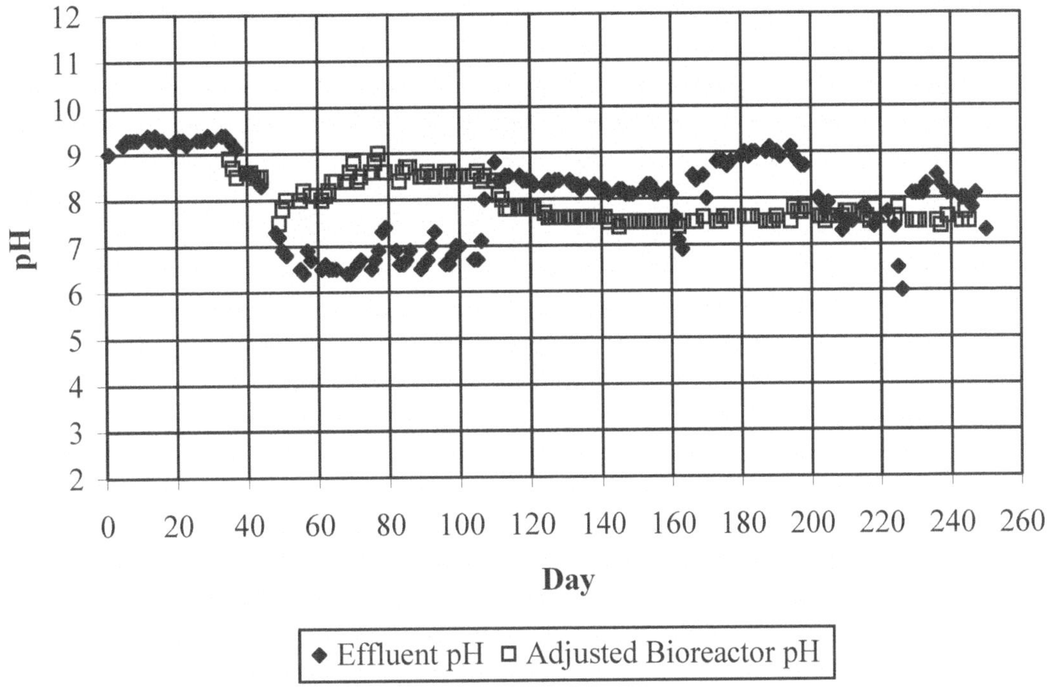

Figure 3-15. Effluent and Adjusted Bioreactor pH; Dewatered Biosolids Supernatant.

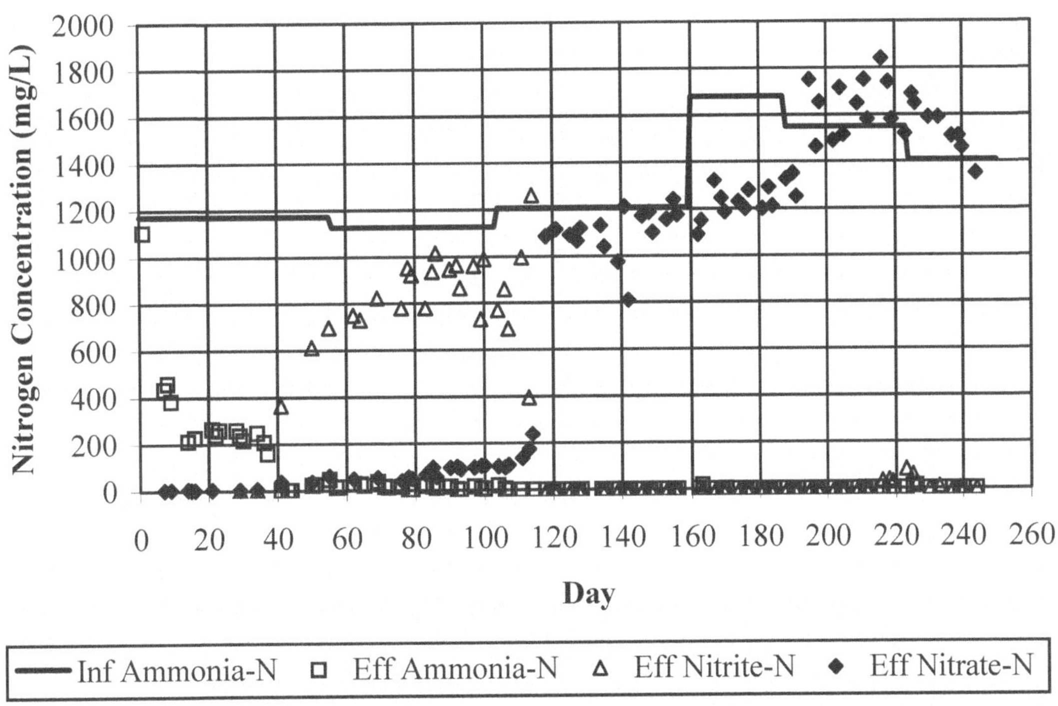

Figure 3-16. Nitrogen Concentrations; Dewatered Biosolids Supernatant (Inf, influent; Eff, effluent).

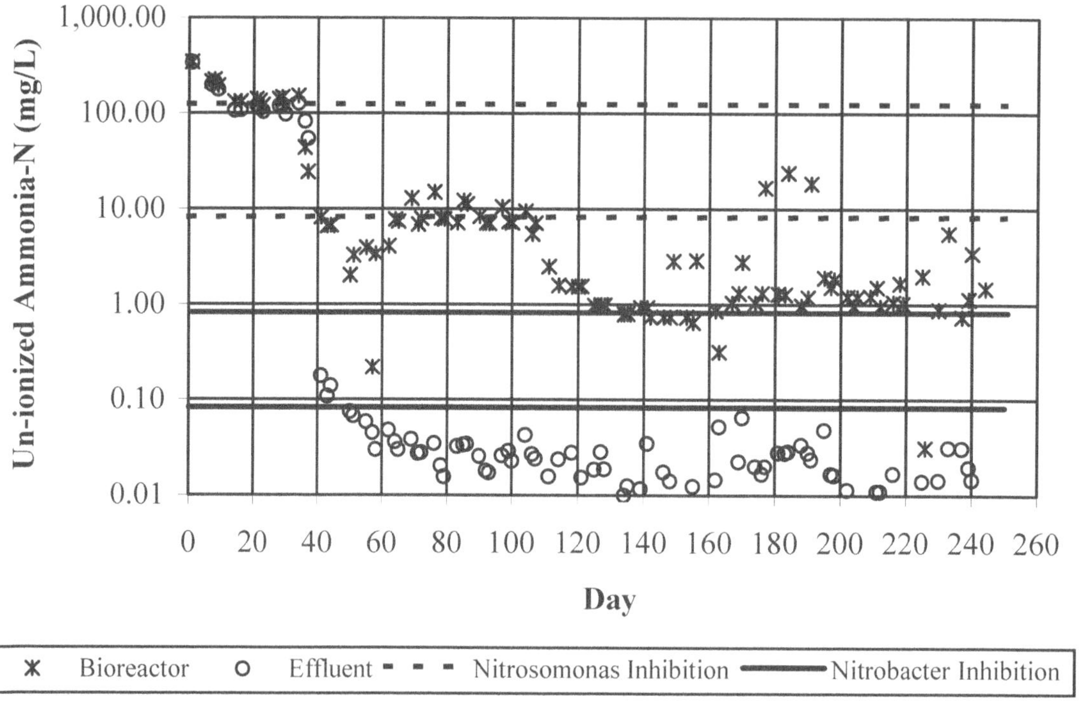

Figure 3-17. Un-ionized Ammonia-nitrogen Concentrations; Dewatered Biosolids Supernatant.

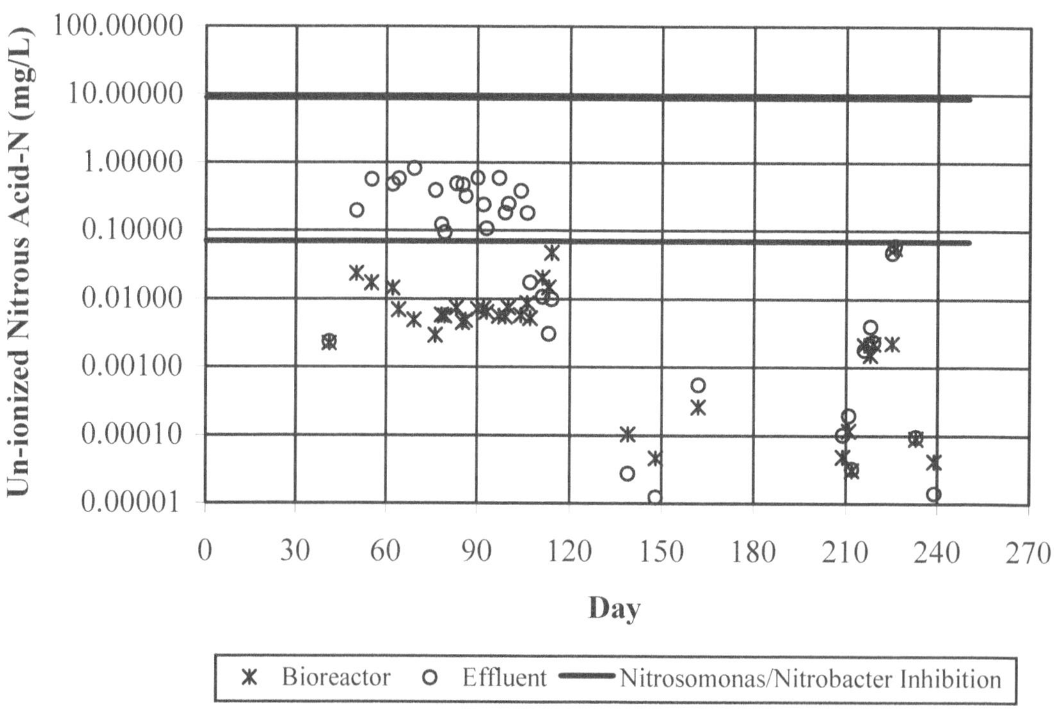

Figure 3-18. Un-ionized Nitrous Acid Concentrations; Dewatered Biosolids Supernatant.

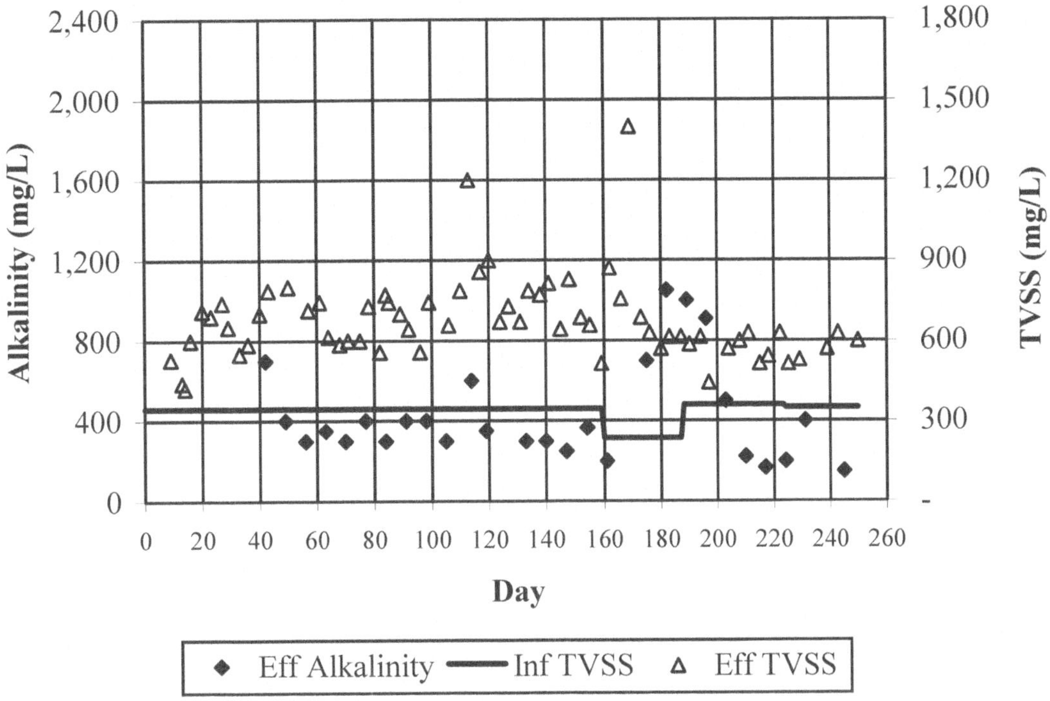

Figure 3-19. Alkalinity and TVSS; Dewatered Biosolids Supernatant (Inf, influent; Eff, effluent).

3.3 Kinetic Evaluation of Seed Bioreactor Performance

Nitrification kinetics for the seed bioreactors used to develop nitrifier populations for subsequent bioaugmentation were evaluated using the specific growth rate (μ_N), ammonium oxidation rate (q_N), and yield coefficient (Y_N). An observed yield coefficient was computed based on the total reactor VSS concentration as follows:

$$Y_{Nobs} = \frac{X}{N_0 - N_1} \quad (3.3\text{-}1)$$

where: Y_{Nobs} = observed yield coefficient, g VSS/g NH_3-N removed
X = reactor VSS, mg/L
N_0 = initial (after dilution in the bioreactor) NH_3-N concentration, mg/L
N_1 = final NH_3-N concentration, mg/L

The reactor active nitrifier concentration and yield coefficient for nitrifiers were estimated as follows:

$$X_{Nest} = X - X_0 \quad (3.3\text{-}2)$$

where: X_{Nest} = estimated reactor active nitrifier concentration (VSS_N), mg/L
X_0 = influent VSS, mg/L

$$Y_{Nest} = \frac{X_{Nest}}{N_0 - N_1} \quad (3.3\text{-}3)$$

where: Y_{Nest} = estimated yield coefficient, g VSS_N grown/g NH_3-N removed

Using the total reactor VSS concentration (X), an observed ammonium oxidation rate (q_{Nobs}) and specific growth rate (μ_{Nobs}) were computed from equations 1.2-6 and 1.2-2. Using the estimated nitrifier concentration (X_{Nest}), an estimated ammonium oxidation rate (q_{Nest}) and specific growth rate (μ_{Nest}) were computed from equations 1.2-6 and 1.2-2 in a similar manner. In the seed bioreactors fed with synthetic waste, the effluent VSS concentration was assumed to be nitrifying biomass. Therefore, Y_{Nobs} and q_{Nobs} were not computed. These parameters were only applicable to the dewatered biosolids supernatant seed reactor with VSS concentrations present that were not associated with nitrifying biomass.

From the data shown in Figures 3-2, 3-6, 3-11, and 3-16, the approximate period of steady-state or quasi-steady-state operation was selected. Average daily constituent concentrations for those periods were used, as described above, to compute kinetic parameters. A summary of these parameters is shown in Table 3-1.

As indicated in Table 3-1, the dewatered biosolids supernatant contained a significant concentration of influent VSS. As a result, the reactor VSS concentration is not a direct representation of the estimated nitrifier concentration. The estimated yield coefficient and ammonium oxidation rate are expected to be a more representative measure of the nitrification kinetics. The estimated yield values for all of the bioreactors fall within typical values reported in the literature. In addition, the estimated ammonium oxidation rate increased with decreasing SRT. However, it should be noted that complete conversion of ammonium to nitrate was not observed in the bioreactors operating at lower SRT values (i.e., 5-day and 10-day SRT). Un-ionized ammonia and un-ionized nitrous acid inhibition are believed to be significant factors for the operational conditions evaluated.

A primary assumption for the successful development of nitrifying microorganisms for subsequent bioaugmentation was that complete, or near complete, nitrification of a high-strength ammonia waste could be achieved. Complete nitrification would be characterized by near total

Table 3-1. Kinetic Parameters for Bioreactor Performance from Average Daily Data.

Parameter	Synthetic waste			Supernatant
	5-day SRT	10-day SRT	20-day SRT	20-day SRT
Approximate period of steady state	Day 35-65	Day 146-252	Day 163-260	Day 118-250
Influent NH_3-N, mg/L	1,005	989	996	1,440
Influent VSS, mg/L	9.0	13.2	13.5	327
Effluent NH_3-N, mg/L	736	190	15.7	1.1
Effluent VSS (X), mg/L	35.7	92.5	148	670
X_{Nest}, mg/L	35.7	92.5	148	343
Y_{Nobs}, g VSS/g NH_3-N removed	N/A	N/A	N/A	0.47
q_{Nobs}, g NH_3-N oxidized/g VSS per day	N/A	N/A	N/A	0.11
μ_{Nobs}, d^{-1}	N/A	N/A	N/A	0.05
Y_{Nest}, g VSS/g NH_3-N removed	0.13	0.12	0.15	0.24
q_{Nest}, g NH_3-N oxidized/g VSS per day	1.51	0.86	0.33	0.21
μ_{Nest}, d^{-1}	0.20	0.10	0.05	0.05

conversion of ammonia to nitrate, and therefore, require that the biological culture contain both ammonia oxidizing and nitrite oxidizing microorganisms. The results of this study indicate that conversion to nitrate was achieved at higher SRT values exhibiting steady-state operation. However, steady-state operation was also observed to depend on the concentrations of ammonia, un-ionized ammonia, and/or un-ionized nitrous acid that were affected by the start-up and operational conditions associated with each SRT (in particular, start-up with high-strength ammonia waste versus low strength with gradual increase to high-strength and batch feed operation). Lower SRTs may be able to achieve stable, complete nitrification under different start-up and operational conditions. These data suggest a potential difficulty in maintaining a stable nitrite oxidizing population of microorganisms with lower SRT values. This potential difficulty must be considered in the design and operation of full-scale system.

For start-up with high-strength ammonia waste, a batch feed process, and low SRT, the results obtained indicate that inhibitory conditions and sequential ammonia and nitrite oxidation were observed. This observation is consistent with recent research conducted by other investigators (Noto et al., 1998; Surmacz-Gorska et al., 1997; Mossakowska et al., 1997). The observed cycling between ammonia removal, nitrite production and conversion, and nitrate production indicates the sensitive balance between *Nitrosomonas* and *Nitrobacter* relative to un-ionized ammonia, un-ionized nitrous acid, and pH. Although it may be possible to culture a population capable of complete nitrification at lower SRTs, the design configuration and operational strategy must consider, mitigate, and/or eliminate the potential for un-ionized ammonia and un-ionized nitrous acid inhibition.

CHAPTER 4.0

BIOAUGMENTATION WITH SEPARATELY CULTURED NITRIFYING MICROORGANISMS

4.1 Bioaugmentation with Nitrifiers Developed from High-strength Ammonia Synthetic Waste at 20-day SRT

Kinetic parameters describing nitrification in the seed bioreactors used to develop supplemental nitrifier populations for subsequent bioaugmentation were evaluated on the basis of average daily data from routine monitoring (as shown in Table 3-1). In addition, more detailed data was obtained during the course of one day of operation, as shown in Figure 4-1. As indicated, ammonia oxidation occurred rapidly in the seed reactor with complete ammonia removal in less than 10 hr. Nitrite accumulation was observed for up to 6 hr, after which nitrite was removed completely. Changes in the nitrate concentration were difficult to analyze precisely because of the high concentration of nitrite when compared with the magnitude of the anticipated increase. At the 20-day SRT, all nitrification was complete in less than 24 hr. The estimated ammonium oxidation rate (q_{Nest}) computed from this data is shown in Table 4-1.

Bioaugmentation analyses were conducted in low-strength ammonia synthetic waste, clarified activated sludge, and non-nitrifying activated sludge. Results from the analyses utilizing nitrifiers developed from high-strength ammonia synthetic waste at a 20-day SRT are summarized in Table 4-1 along with the data from Run #1 conducted in the seed bioreactor without supplemental nitrifiers. Nitrogen concentrations for each test run are shown in Figures 4-2 through 4-6. Results from selected oxygen uptake analyses are shown in Figure 4-7.

Table 4-1. Bioaugmentation Analyses with Nitrifiers from High-strength Ammonia Synthetic Waste at 20-day SRT.

Run #	1	2	3	4	5
Substrate	Seed bioreactor	Low-strength synthetic waste		Clarified activated sludge	Activated sludge
Substrate volume, L	N/A	5.75	3.50	3.50	3.50
Supplemental nitrifier volume, L	N/A	0.25	0.50	0.50	0.50
Total volume, L	5.00	6.00	4.00	4.00	4.00
X_{Nest} (VSS_N), mg/L	109	6.8	15.4	14.2	15.0
NH_3-N removal rate, mg/L per hr	5.78	0.66	0.72	0.94	0.86
q_{Nest}, mg NH_3-N oxidized/mg VSS_N per day	1.28	2.34	1.12	1.59	1.38
Oxygen uptake rate, mg/L per hr	N/A	N/A	2.97	2.96	10.59
Specific oxygen uptake, mg O_2/mg NH_3-N oxidized	N/A	N/A	4.13	3.15	N/A

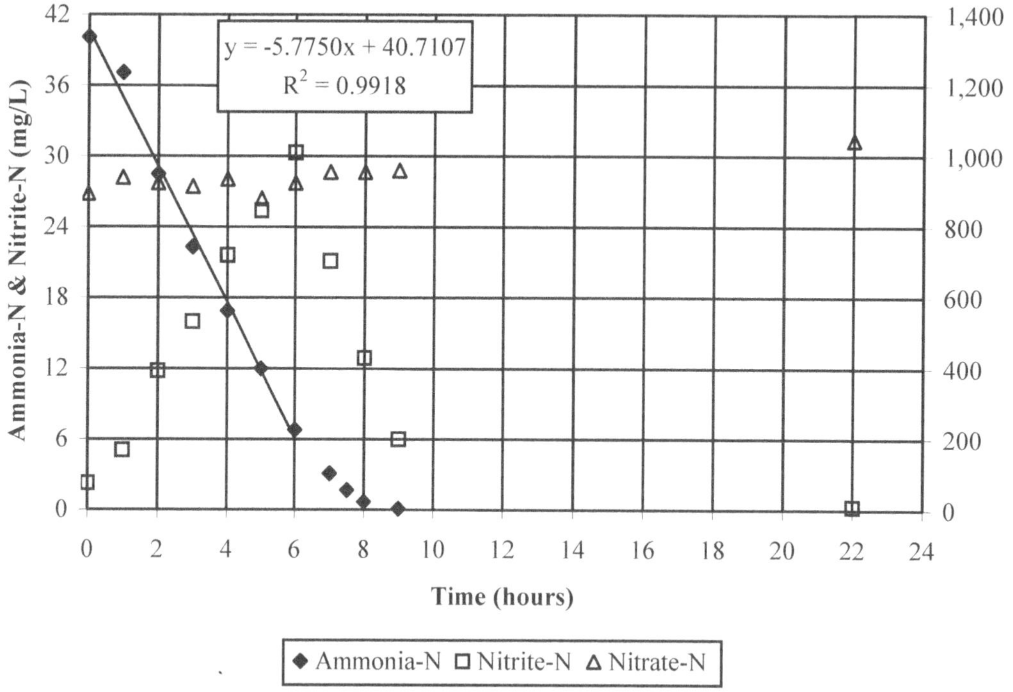

Figure 4-1. Run #1; Seed Bioreactor Nitrogen Concentrations (20.5°C, 8.1 S.U.).

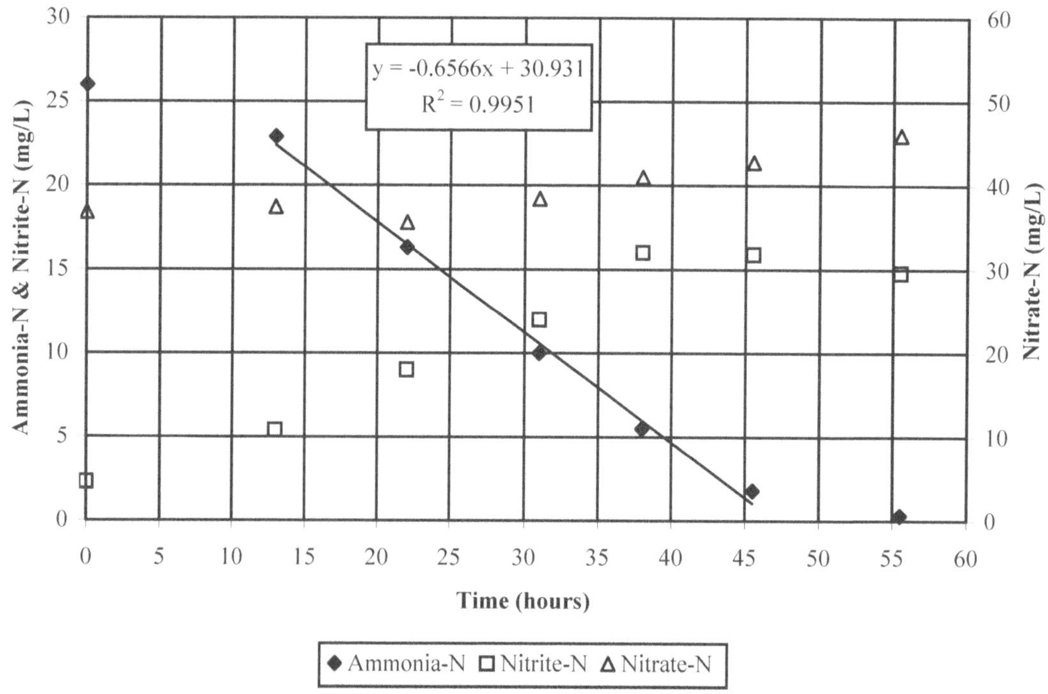

Figure 4-2. Run #2; Bioaugmentation in Low-strength Synthetic Waste with Nitrifiers Cultured in High-strength Synthetic Waste (22.2°C, 7.7 S.U.).

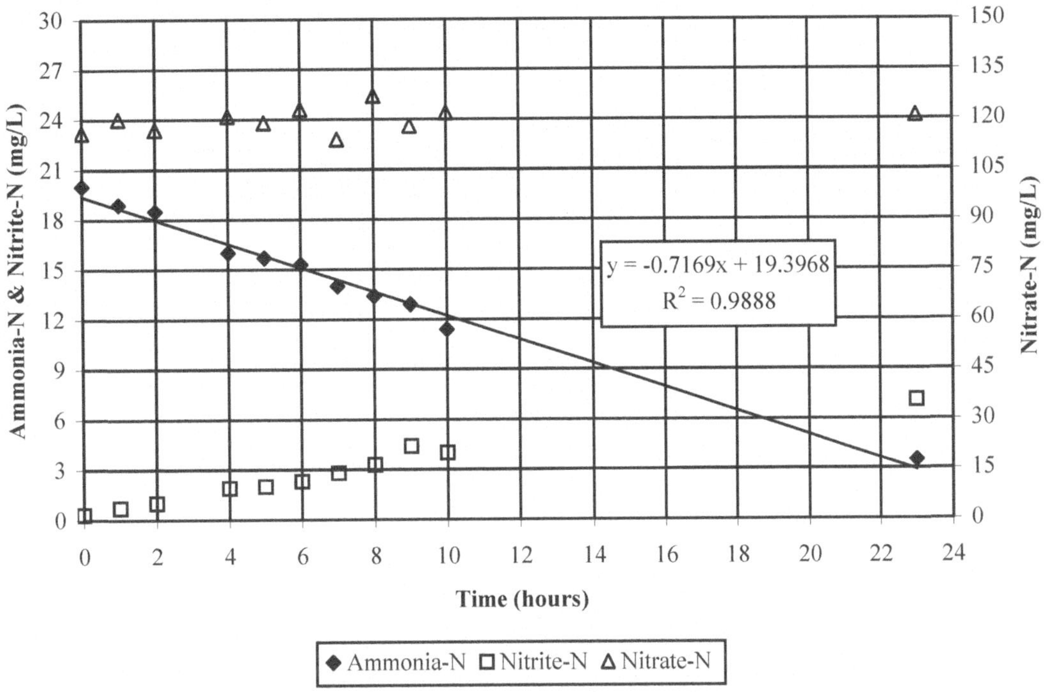

Figure 4-3. Run #3; Bioaugmentation in Low-strength Synthetic Waste with Nitrifiers Cultured in High-strength Synthetic Waste (20.2°C, 7.5 S.U.).

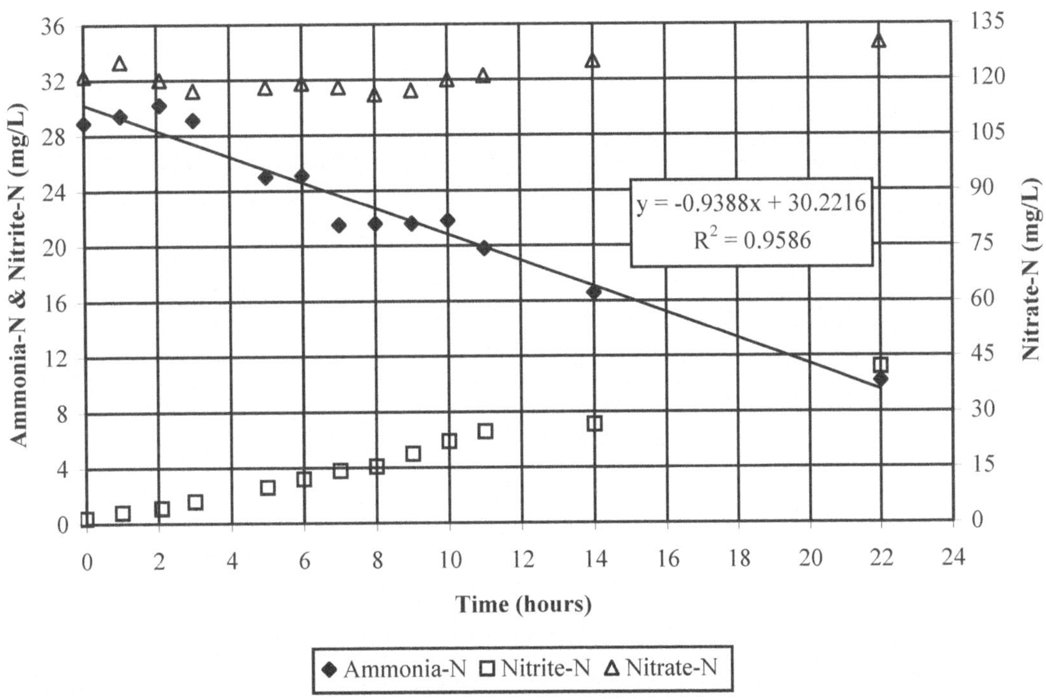

Figure 4-4. Run #4; Bioaugmentation in Clarified Activated Sludge with Nitrifiers Cultured in High-strength Synthetic Waste (19.3°C, 7.6 S.U.).

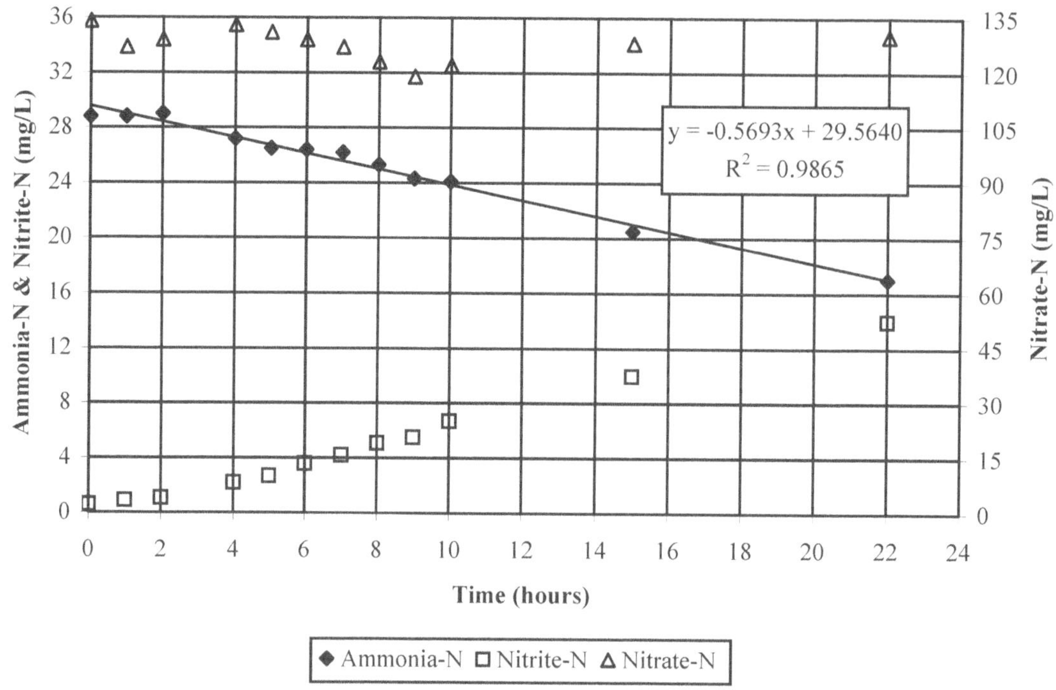

Figure 4-5. Run #5; Bioaugmentation in Activated Sludge with Nitrifiers Cultured in High-strength Synthetic Waste (18.3°C, 7.2 S.U.).

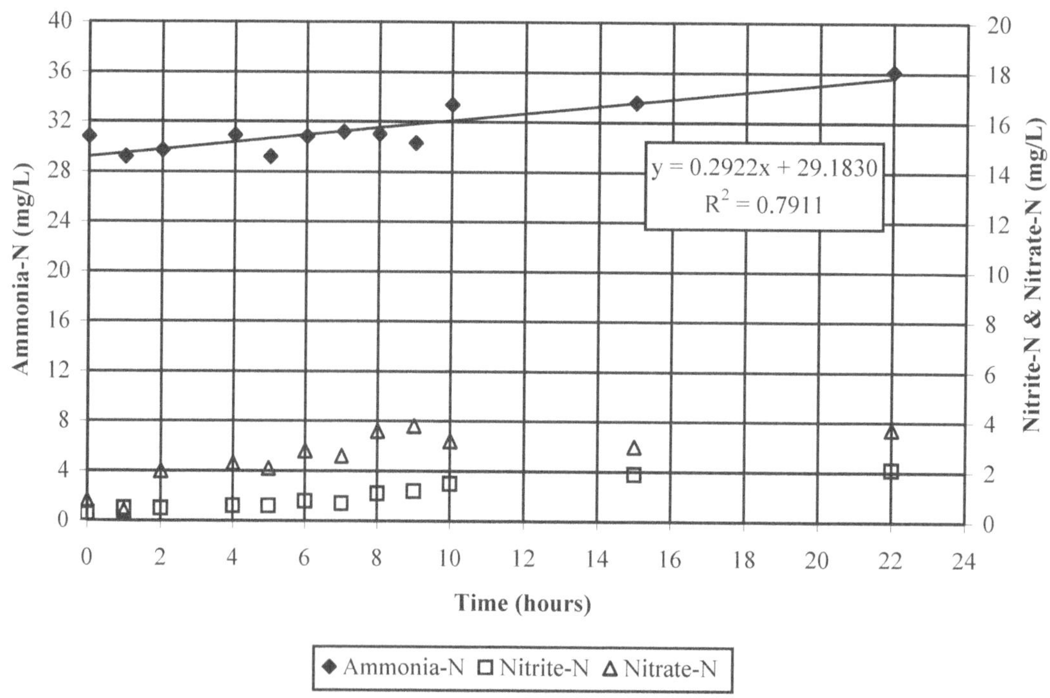

Figure 4-6. Run #5 (Control); Bioaugmentation in Activated Sludge with Nitrifiers Cultured in High-strength Synthetic Waste (18.2°C, 7.2 S.U.).

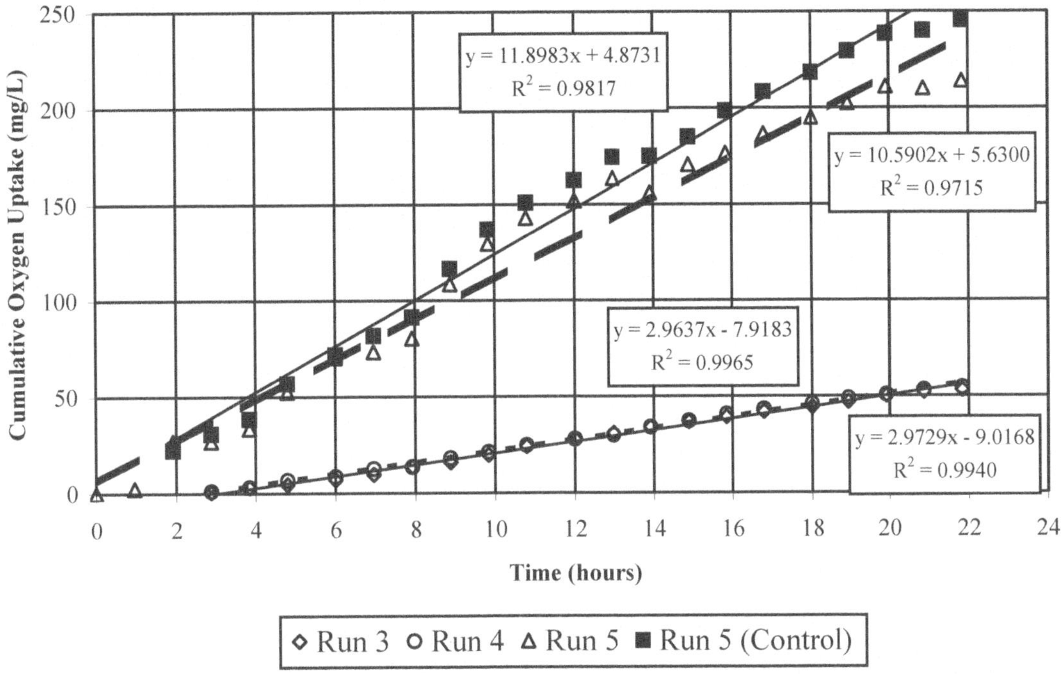

Figure 4-7. Oxygen Uptake Analyses for Bioaugmentation in Low-strength Synthetic Waste (Run #3), Clarified Activated Sludge (Run #4), and Activated Sludge (Run #5) with Nitrifiers Cultured in High-strength Synthetic Waste.

Runs #2 and #3 were conducted to verify bioaugmentation procedures by subjecting the supplemental nitrifier population to a familiar substrate, a low-strength ammonia synthetic waste at similar temperature and pH. As indicated in Figures 4-2 and 4-3, no lag or acclimation period was apparent in ammonia removal and nitrite production. The ammonium oxidation rates (q_{Nest}) for each run were not significantly different from the ammonium oxidation rate observed in the seed reactor (Run #1) but were significantly greater than 0.33 mg NH_3-N oxidized/mg VSS_N per day; the rate computed for the seed bioreactor from daily average data (Table 3-1). Therefore, with no discernable acclimation period being observed, bioaugmentation procedures were verified.

Runs #4 and #5 were conducted to determine if an acclimation period was required when the separately cultured nitrifier population was introduced to an unfamiliar substrate as would be the case when used to bioaugment a non-nitrifying activated sludge. As indicated in Figures 4-4 and 4-5, ammonia removal and nitrite production were observed with no significant lag or acclimation period.

The intent of the control for Run #5 was to account for two possible processes; the ammonification of organic-nitrogen constituents that may be present in these substrates and the biological uptake of nitrogen by non-nitrifying biomass. The control for Run #5 confirmed that the net result of these two processes was the additional release of ammonia. This information was used to compute q_{Nest} for this test run.

The ammonium oxidation rates (q_{Nest}) for Runs #4 and #5 were both greater than the ammonium oxidation rate observed in the seed reactor (Run #1) and significantly greater than 0.33 mg NH_3-N oxidized/mg VSS_N per day; the rate computed for the seed bioreactor from daily average data (Table 3-1). Therefore, it can be concluded that no significant acclimation period was observed and that the nitrifiers were able to quickly adapt to the unfamiliar substrate.

In all of the bioaugmentation test runs the initial removal of ammonia was a conversion to nitrite with little, if any, conversion to nitrate. It would appear that conditions more favorable to *Nitrosomonas* exist under the initial conditions and that, because the ammonia concentration is reduced, the oxidation of nitrite to nitrate can begin to take place.

The oxygen uptake data for Runs #3 and #4, where oxygen use associated with nitrification can be more accurately evaluated, is consistent with the observations from Figures 4-3 and 4-4. The specific oxygen uptake observed was less than the stochiometric requirement for complete nitrification (4.6 mg O_2/mg NH_3-N oxidized) for both test runs. This observation coincides with the incomplete nitrification that was observed based on nitrogen constituents. The oxygen uptake rate associated with nitrification for Run #5 (with activated sludge) is not discernable because of the high oxygen use associated with the heterotrophic biomass.

For all of the bioaugmentation analyses, the temperature and pH of the seed bioreactor and bioaugmentation bioreactor were not significantly different. The temperature difference was within ±2.1°C and the pH difference was within ±0.5 S.U. Nitrifiers are known to be sensitive to these environmental conditions. If supplemental nitrifiers are added to an activated sludge system with significantly different temperature and/or pH, some acclimation to this environmental stress could be expected. This may range from total inhibition to reduced nitrification rates (i.e., caused by lower temperature and/or pH), thereby requiring that a greater quantity of supplemental nitrifiers be added to achieve the desired nitrification in the activated sludge system.

4.2 Bioaugmentation with Nitrifiers Developed from Dewatered Biosolids Supernatant at 20-day SRT

More detailed data for the nitrification kinetics in the dewatered biosolids supernatant seed bioreactor were obtained during the course of one day of operation to compare to that computed from average daily data from routine monitoring (Table 3-1). As indicated in Figure 4-8, ammonia removal was essentially complete within 12 hr. Nitrite accumulation was noted through the first 11 hr. Nitrite had been removed completely after 22 hr. Changes in the nitrate concentration were difficult to analyze because of the high concentration. At the 20-day SRT, all nitrification was complete in less than 24 hr.

Because of the presence of VSS in the dewatered biosolids supernatant not associated with the nitrifying biomass, observed and estimated ammonium oxidation rates were computed. The observed ammonium oxidation rate (q_{Nobs}) is based on the total VSS concentration; the estimated ammonium oxidation rate (q_{Nest}) is based on the estimated VSS concentration associated with the nitrifying biomass computed as described in Section 3.3. Seed bioreactor q_{Nobs} and q_{Nest} computed from data in Figure 4-8 are shown in Table 4-2.

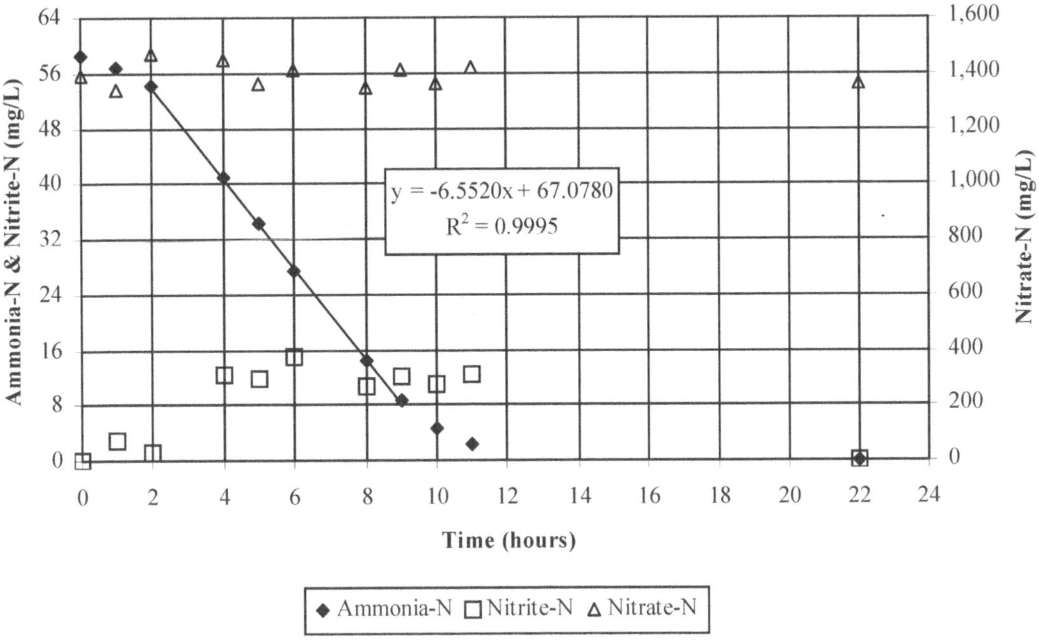

Figure 4-8. Run #6; Seed Bioreactor Nitrogen Concentrations (18.9°C, 7.7 S.U.).

Bioaugmentation analyses were conducted in diluted supernatant, clarified activated sludge, and non-nitrifying activated sludge. Results from the analyses using nitrifiers developed from dewatered biosolids supernatant at a 20-day SRT are summarized in Table 4-2. Also included in Table 4-2 is data from Run #6 conducted in the seed bioreactor without supplemental nitrifiers. Nitrogen concentrations for each test run are shown in Figures 4-9 through 4-14. Results from selected oxygen uptake analyses are shown in Figure 4-15.

Table 4-2. Bioaugmentation Analyses with Nitrifiers from Dewatered Biosolids Supernatant at 20-day SRT.

Run #	6	7	8	9	10	11
Substrate	Seed bioreactor	Diluted dewatered supernatant		Clarified activated sludge		Activated sludge
Substrate volume, L	N/A	4.75	3.55	4.75	3.55	3.55
Supplemental nitrifier volume, L	N/A	0.25	0.45	0.25	0.45	0.45
Total volume, L	5.00	5.00	4.00	5.00	4.00	4.00
X (VSS), mg/L	558	34.5	59.0	31.3	61.4	77.2
X_{Nest} (VSS_N), mg/L	285	17.6	30.1	16.0	31.3	39.4
NH_3-N removal rate, mg/L per hr	6.55	0.85	0.62	0.64	0.72	0.65
q_{Nobs}, mg NH_3-N oxidized/mg VSS per day	0.28	0.59	0.25	0.49	0.28	0.20
q_{Nest}, mg NH_3-N oxidized/mg VSS_N per day	0.55	1.16	0.50	0.97	0.55	0.40
Oxygen uptake rate, mg/L per hr	N/A	N/A	2.32	N/A	2.97	10.74
Specific oxygen uptake, mg O_2/mg NH_3-N oxidized	N/A	N/A	3.74	N/A	4.13	N/A

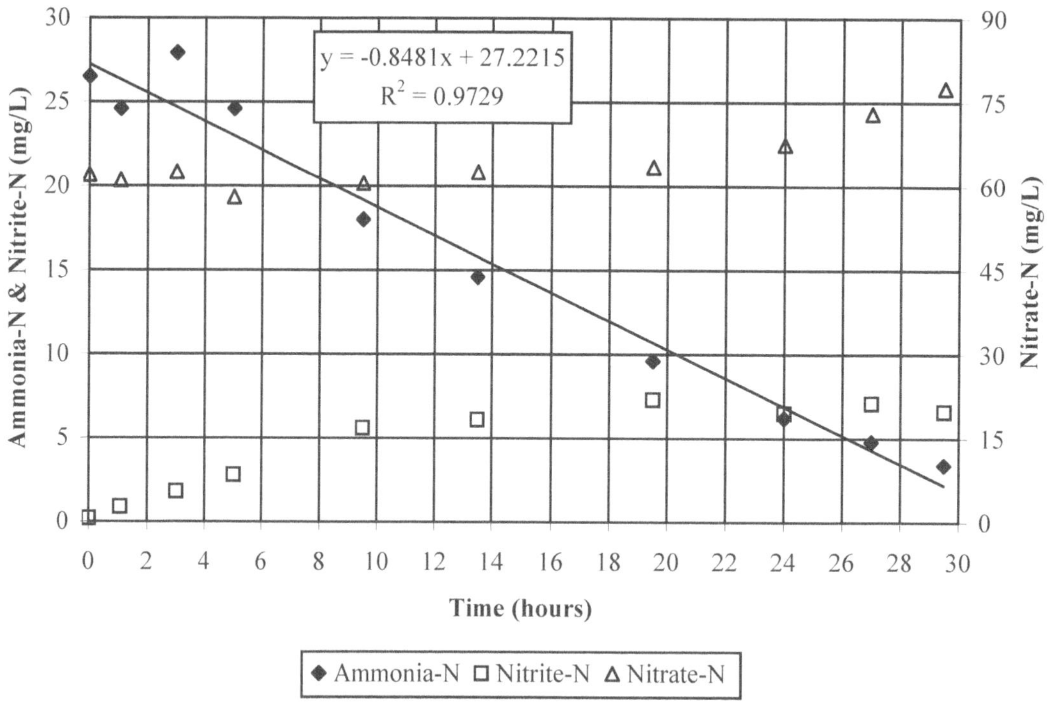

Figure 4-9. Run #7; Bioaugmentation in Diluted Supernatant with Nitrifiers Cultured in Dewatered Biosolids Supernatant (22.4°C, 7.7 S.U.).

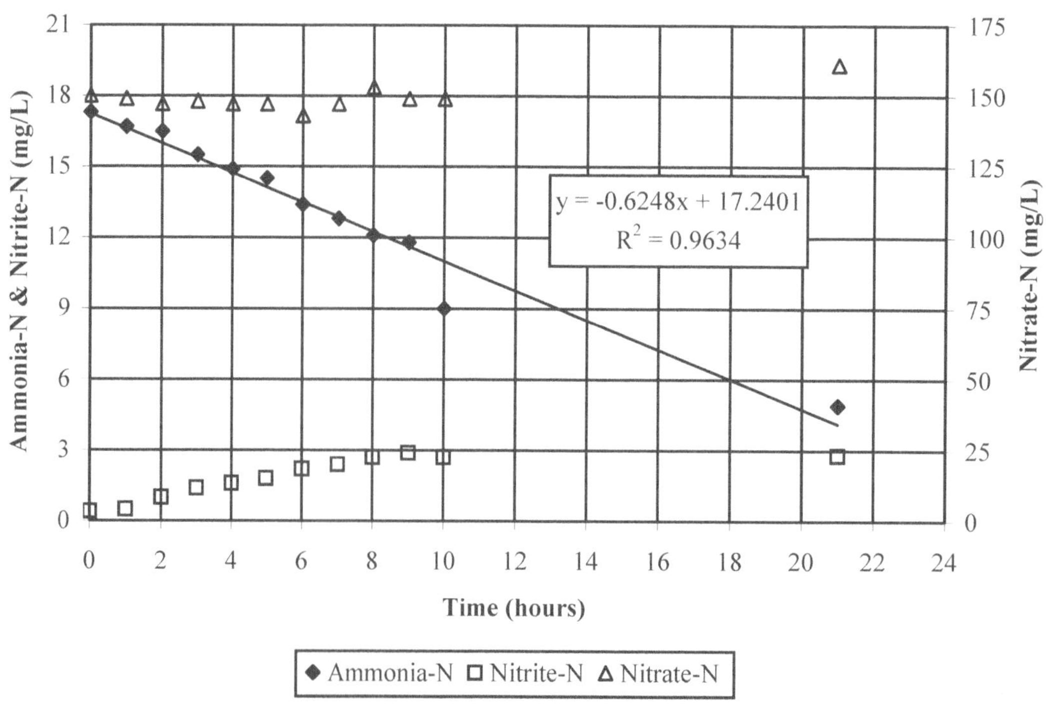

Figure 4-10. Run #8; Bioaugmentation in Diluted Supernatant with Nitrifiers Cultured in Dewatered Biosolids Supernatant (20.5°C, 7.7 S.U.).

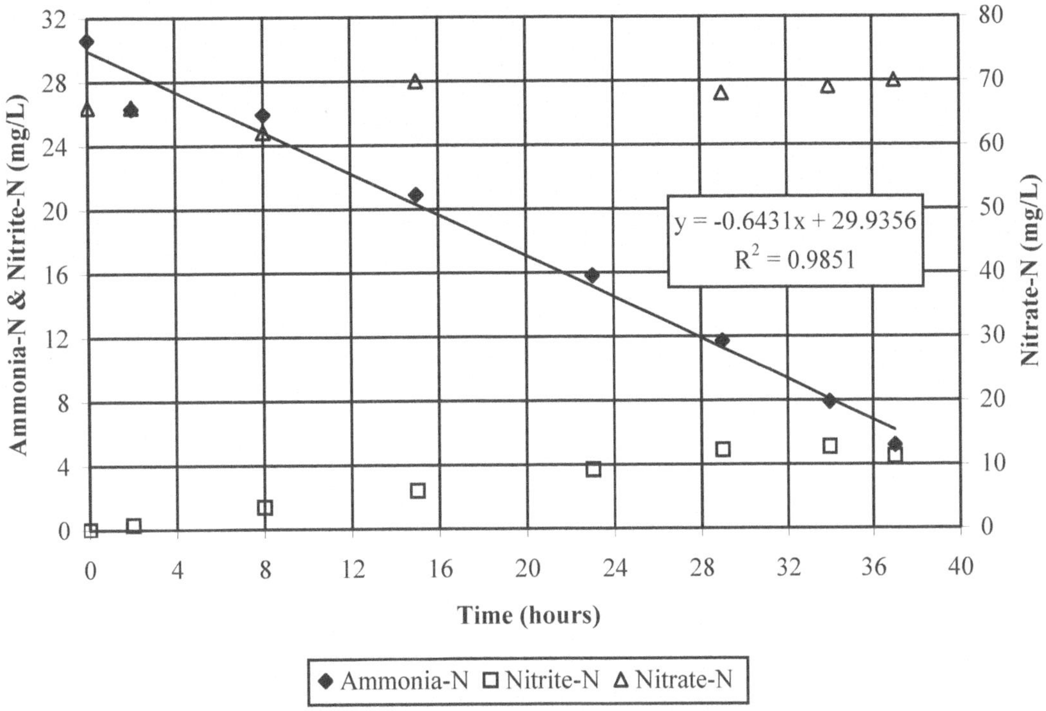

Figure 4-11. Run #9; Bioaugmentation in Clarified Activated Sludge with Nitrifiers Cultured in Dewatered Biosolids Supernatant (23.8°C, 7.6 S.U.).

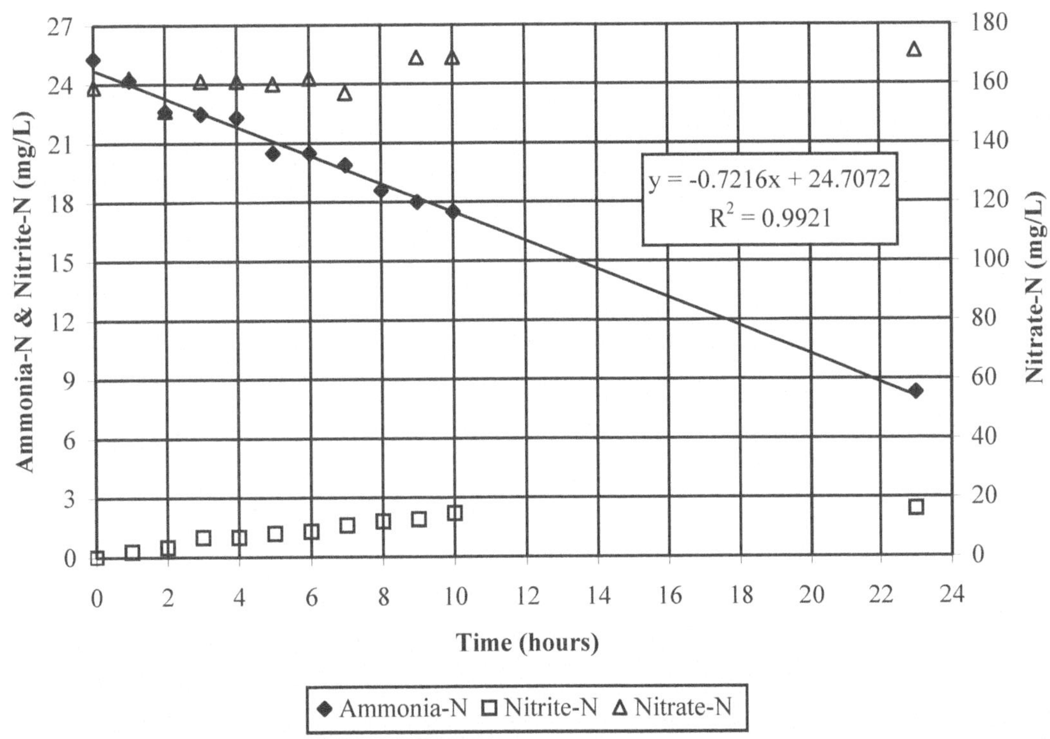

Figure 4-12. Run #10; Bioaugmentation in Clarified Activated Sludge with Nitrifiers Cultured in Dewatered Biosolids Supernatant (18.7°C, 7.9 S.U.).

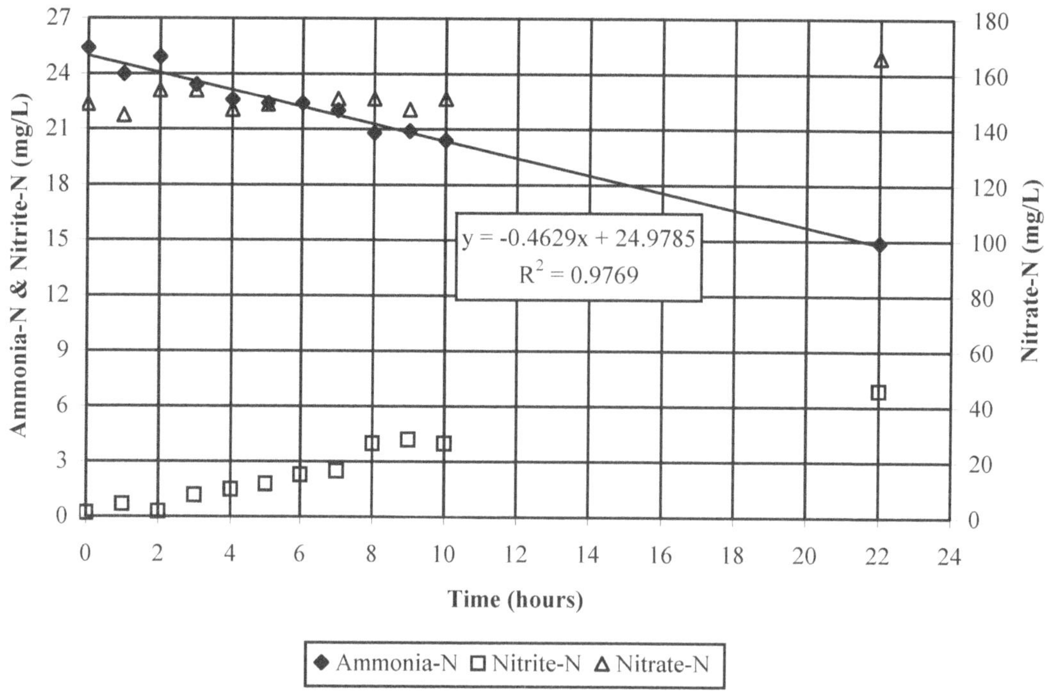

Figure 4-13. Run #11; Bioaugmentation in Activated Sludge with Nitrifiers Cultured in Dewatered Biosolids Supernatant (17.9°C, 7.2 S.U.).

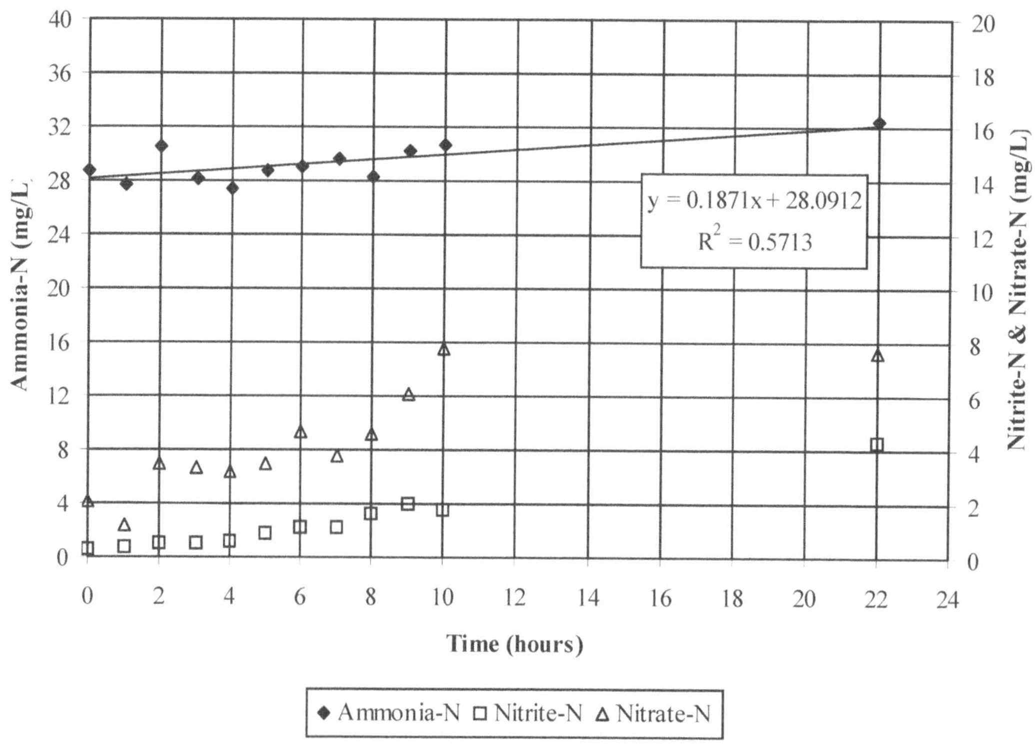

Figure 4-14. Run #11 (Control); Bioaugmentation in Activated Sludge with Nitrifiers Cultured in Dewatered Biosolids Supernatant (17.7°C, 7.0 S.U.).

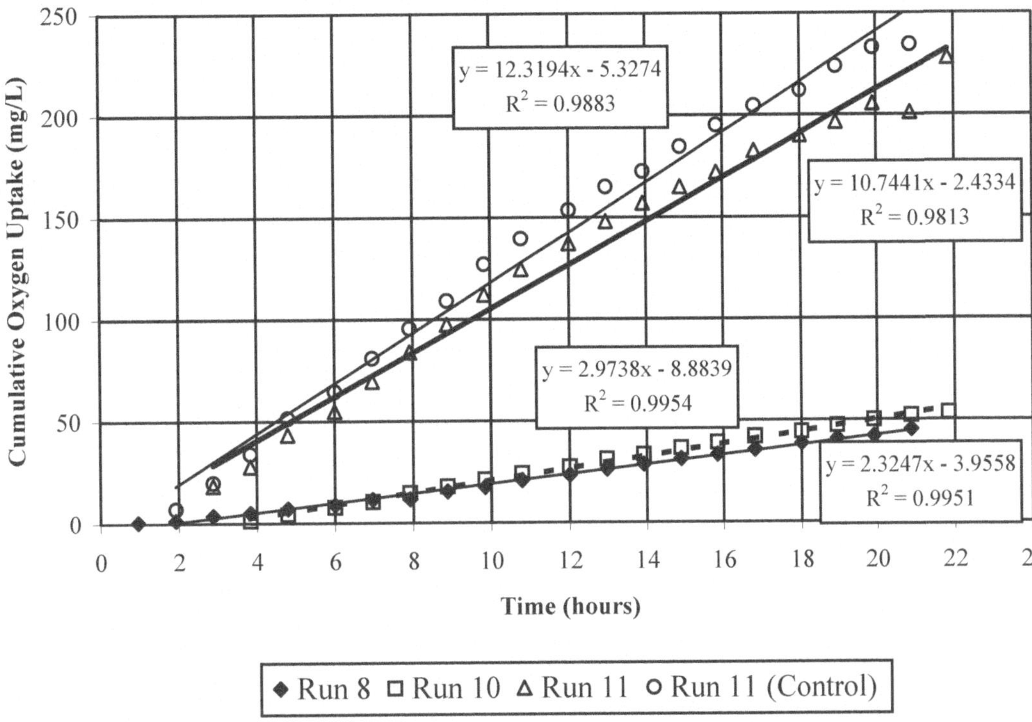

Figure 4-15. Oxygen Uptake Analyses for Bioaugmentation in Diluted Supernatant (Run #8), Clarified Activated Sludge (Run #10), and Activated Sludge (Run #11) with Nitrifiers Cultured in Dewatered Biosolids Supernatant.

Runs #7 and #8 were conducted to verify bioaugmentation procedures by subjecting the supplemental nitrifier population to a familiar substrate, diluted dewatered biosolids supernatant. As indicated in Figures 4-9 and 4-10, no lag or acclimation period was apparent. The observed and estimated ammonium oxidation rates for Run #7 were both significantly greater than that for the seed bioreactor (Run #6). The observed and estimated ammonium oxidation rates were only slightly less than the rates for the seed bioreactor for Run #8. The rates computed for both runs were significantly greater than those computed for the seed bioreactor from average daily data with observed and estimated ammonium oxidation rates of 0.11 NH_3-N oxidized/mg VSS per day and 0.21 mg NH_3-N oxidized/mg VSS_N per day, respectively (Table 3-1). Therefore, with no discernable acclimation period being observed, bioaugmentation procedures were verified.

Runs #9 through #11 were conducted to determine if an acclimation period was required when the separately cultured nitrifier population was used with an unfamiliar substrate as would be the case when introduced into non-nitrifying activated sludge. As indicated in Figures 4-11 through 4-13, ammonia removal and nitrite production were observed with no significant lag or acclimation period.

The control for Run #11 confirms that the net result of ammonification of organic nitrogen and biological nitrogen uptake was an additional release of ammonia. This information was used in computing the ammonium oxidation rate for this test run.

The observed and estimated ammonium oxidation rates for Runs #9 and #10 were both equal to, or greater than, those for the seed bioreactor (Run #6). For bioaugmentation in activated sludge (Run #11), the observed and estimated ammonium oxidation rates were approximately 30% below those for the seed bioreactor (Run #6). However, as shown in Figure 4-13, an acclimation period was not apparent. All of the rates for Runs #7 through #11 were greater than the observed and estimated ammonium oxidation rates for the seed bioreactor computed from average daily data (0.11 mg NH_3-N oxidized/mg VSS per day and 0.21 mg NH_3-N oxidized/mg VSS_N per day, respectively). Therefore, it can be concluded that, no significant acclimation period was required.

As was observed for the synthetic waste seed nitrifiers, the initial removal of ammonia was a conversion to nitrite with little, if any, conversion to nitrate. Again, it would appear that conditions more favorable to *Nitrosomonas* exist initially and that, as the ammonia concentration is reduced, the oxidation of nitrite to nitrate can begin to take place.

The oxygen uptake data for Runs #8 and #10, where oxygen use associated with nitrification can be more accurately evaluated, is consistent with the observations from Figures 4-10 and 4-12. The specific oxygen uptake observed was less than the stochiometric requirement for complete nitrification which coincides with the incomplete nitrification that was observed based on nitrogen constituents. The oxygen uptake rate associated with nitrification for Run #11 (with activated sludge) is not discernable because of the high oxygen uptake rate associated with the heterotrophic biomass.

For all of the bioaugmentation analyses, the temperature and pH of the seed bioreactor and bioaugmentation bioreactor were not significantly different with a temperature difference within ±4.5°C and a pH difference within ±0.8 S.U. The potential for differences in environmental conditions between seed and bioaugmentation systems in a full-scale system will depend on the source of the dewatered biosolids supernatant. For example, if biosolids are obtained directly from an anaerobic digestion process, the temperature of the supernatant will typically be greater than that of the activated sludge system. Detention time in the seed nitrification system would reduce this temperature to some degree but it is likely that the seed nitrifiers would be fed into a non-nitrifying activated sludge system operating at a lower temperature. This may result in a decrease in microbial activity (nitrification rate) upon bioagumentation, which would require that a greater quantity of supplemental nitrifiers be added to achieve nitrification in the activated sludge system. Biosolids from secondary digesters or from liquid biosolids storage facilities may have significant temperature variation. These variations are both greater than and less than the activated sludge system, depending on season. These variations would need careful consideration in designing the seed nitrification system.

Seed nitrifiers from dewatered biosolids supernatant may also be developed at a pH different from that of an activated sludge system. This pH could be higher or lower depending on the operation of the seed nitrification system and whether or not chemical addition (for alkalinity) is required. If the pH difference is significant, the microbial activity of the seed nitrifiers in the activated sludge system might increase, decrease, or be inhibited. Reduced nitrification rates because of pH might be overcome by increasing the quantity of seed nitrifiers added. Inhibitory pH conditions may require adjustment in the activated sludge system.

Nitrifiers are known to be sensitive to these environmental conditions. If supplemental nitrifiers are added to an activated sludge system with significantly different temperature and/or pH, some acclimation to this environmental stress could be expected. This may range from total inhibition to reduced nitrification rates (i.e., caused by lower temperature and/or pH) thereby requiring that a greater quantity of supplemental nitrifiers be added to achieve the desired nitrification in the activated sludge system. A multitude of environmental conditions must be considered in process design and operation.

4.3 Bioaugmentation with Nitrifiers from Nitrifying Trickling Filter Biomass

Direct bioaugmentation in low-strength synthetic waste, clarified activated sludge, and non-nitrifying activated sludge was conducted using nitrifying trickling filter biomass. Three different substrates were used to compare kinetic parameters from bioaugmentation analyses because kinetic parameters from seed bioreactors (comparable to Runs #1 and #6) were not available. Results from bioaugmentation analyses are shown in Table 4-3 and Figures 4-16 through 4-23.

Table 4-3. Bioaugmentation Analyses with Nitrifiers from Trickling Filter Biomass.

Run #	13	14	15	16	17 (Duplicate)	18	19
Substrate	Low-strength synthetic Waste		Clarified activated sludge			Activated sludge	
Substrate volume, L	5.730	5.864	5.730	5.864	5.685	5.767	5.883
Supplemental nitrifier Volume, L	0.270	0.136	0.270	0.136	0.315	0.233	0.117
Total volume, L	6.000	6.000	6.000	6.000	6.000	6.000	6.000
X (VSS), mg/L	490	320	490	320	440	528	265
NH_3-N removal rate, mg/L per hr	1.71	0.87	1.71	1.11	1.24 (1.27)	1.46	0.81
q_{Nobs}, mg NH_3-N oxidized/mg VSS per day	0.084	0.066	0.084	0.083	0.067 (0.069)	0.067	0.074

Runs #13 and #14 were conducted in low-strength ammonia synthetic wastewater. Runs #15 through #17 were conducted in clarified effluent from non-nitrifying activated sludge (i.e., final clarifier effluent). Runs #17 and #18 were conducted in non-nitrifying activated sludge. Run #17 was conducted in duplicate. The trickling filter biomass contained VSS concentrations not associated with the actual nitrifying biomass. Because a measure or estimate of the actual concentration of nitrifying biomass cannot be directly obtained, only an observed ammonium oxidation rate (q_{Nobs}) can be computed for these analyses.

As indicated in Figures 4-16 through 4-23, ammonia removal was observed in all bioaugmentation analyses with no apparent lag or acclimation period. Nitrite production was also immediately observed. Although nitrite accumulation was observed to some degree in all test runs, the greatest accumulation (on the basis of maximum nitrite concentration measured) occurred in the activated sludge analyses. Unlike the bioaugmentation analyses with nitrifiers cultured in synthetic wastewater or dewatered biosolids supernatant (Runs #1–#12), nitrate production occurred almost immediately with the trickling filter biomass. This result may be related to how well established the populations of *Nitrosomonas* and *Nitrobacter* were in the trickling filter biomass. These populations are likely to be more stable in the trickling filter biomass obtained from a full-scale operating facility and, as such, may be less susceptible to upset when used for bioaugmentation.

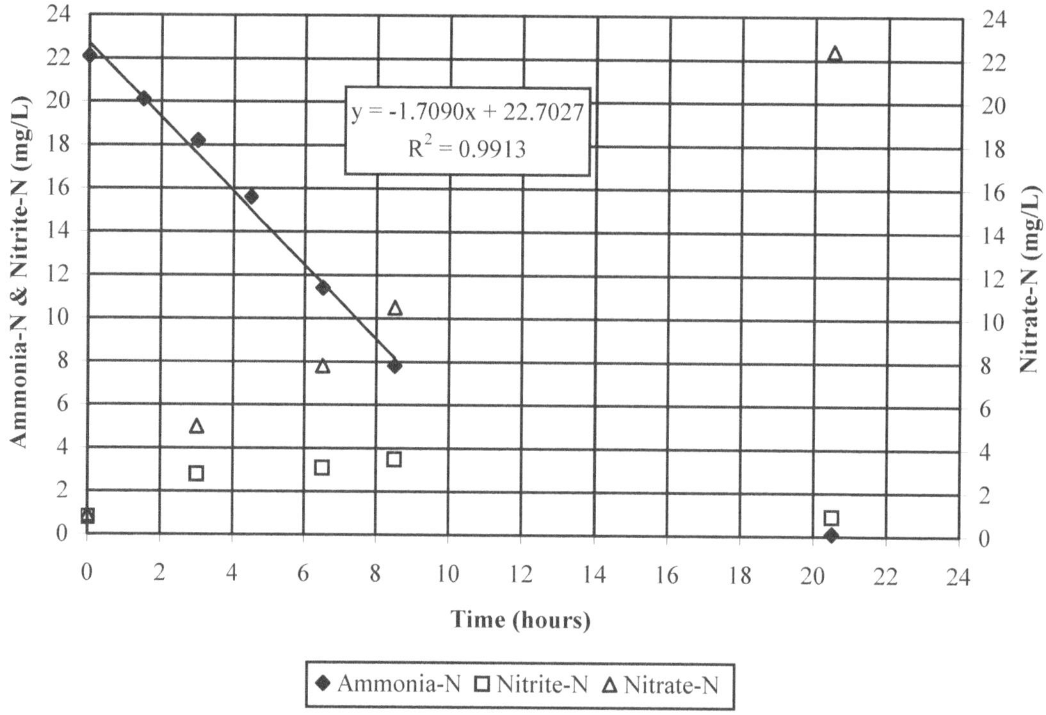

Figure 4-16. Run #13; Bioaugmentation in Low-strength Synthetic Waste with Nitrifying Trickling Filter Biomass (24.2°C, 7.6 S.U.).

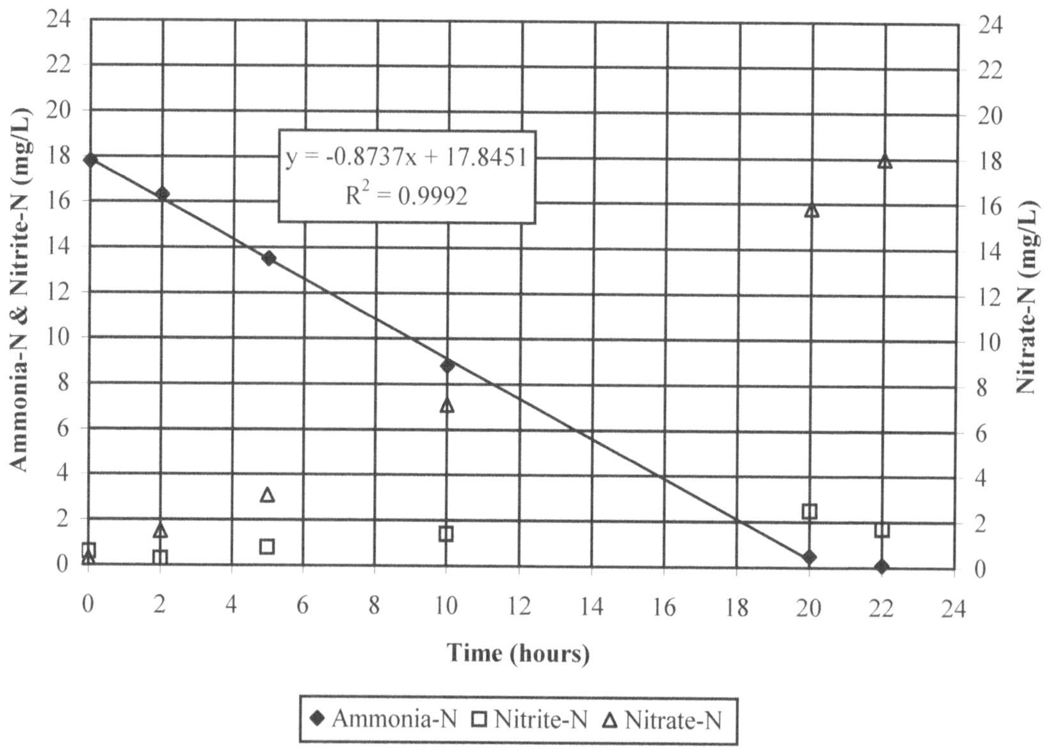

Figure 4-17. Run #14; Bioaugmentation in Low-strength Synthetic Waste with Nitrifying Trickling Filter Biomass (21.2°C, 7.7 S.U.).

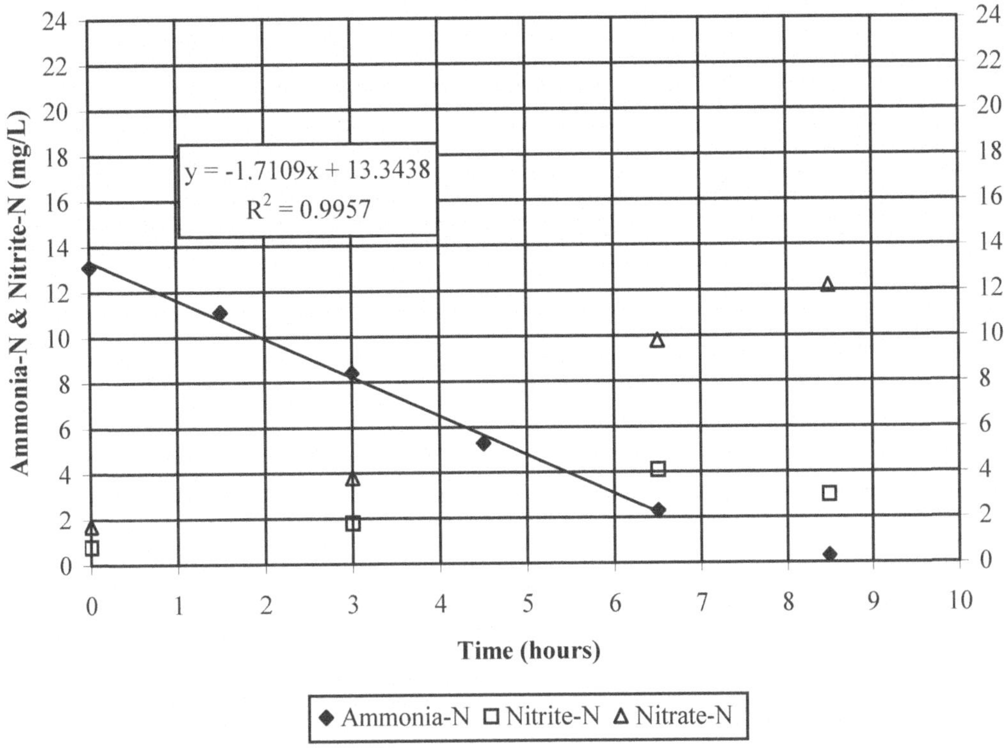

Figure 4-18. Run #15; Bioaugmentation in Clarified Activated Sludge with Nitrifying Trickling Filter Biomass (25.6°C, 7.7 S.U.).

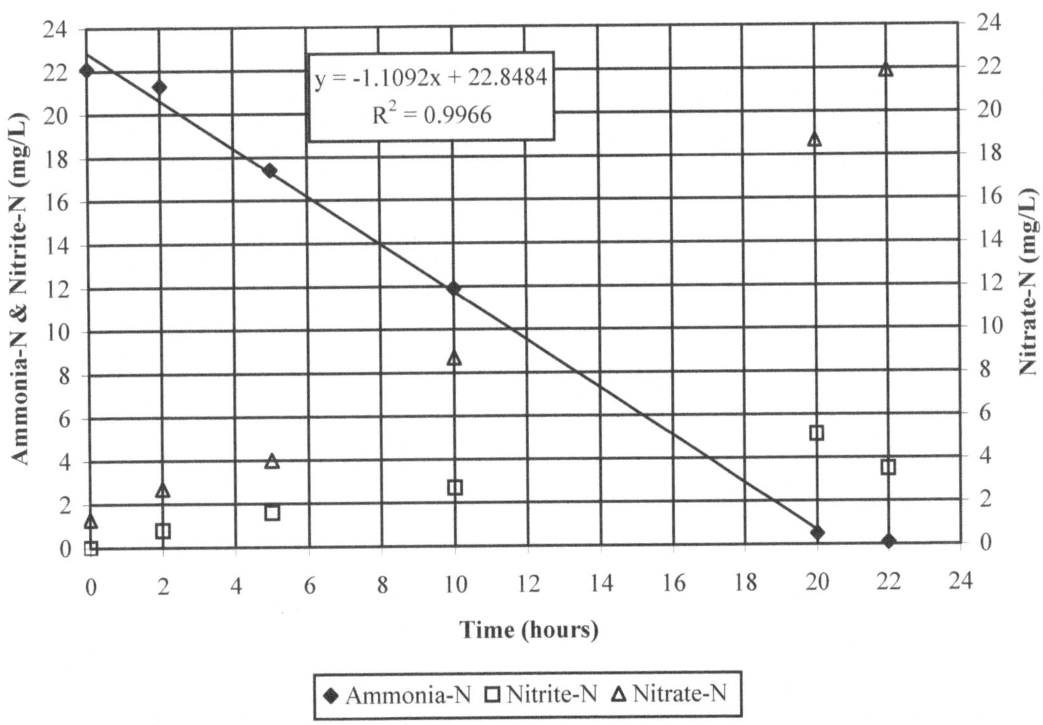

Figure 4-19. Run #16; Bioaugmentation in Clarified Activated Sludge with Nitrifying Trickling Filter Biomass (20.8°C, 8.1 S.U.).

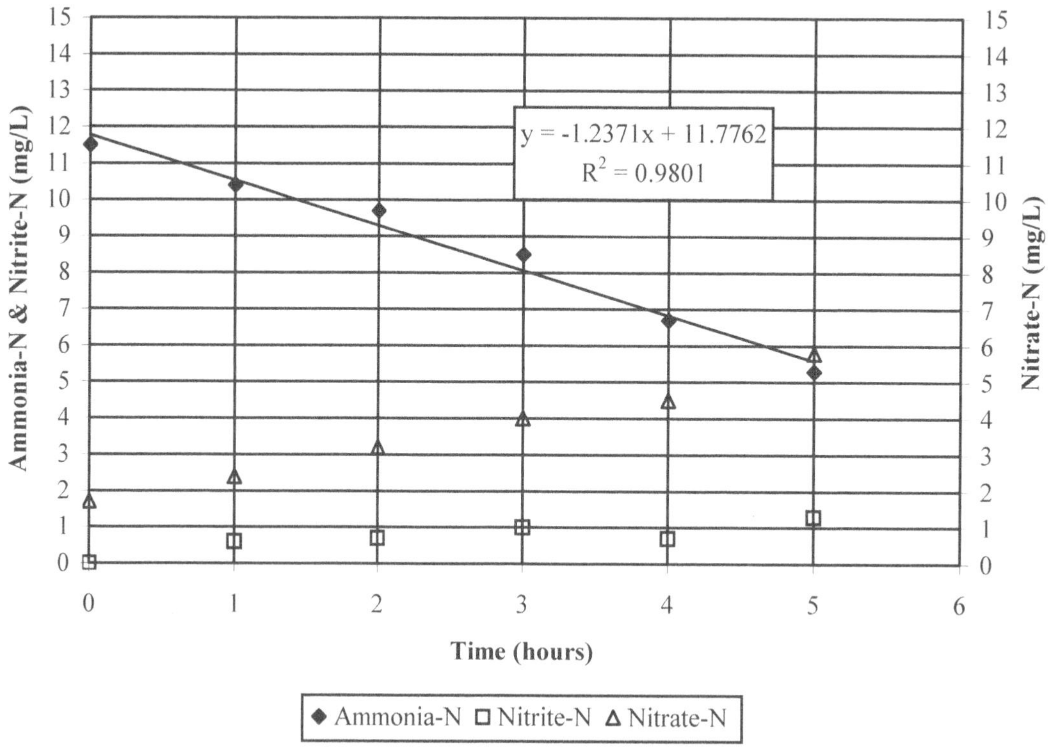

Figure 4-20. Run #17; Bioaugmentation in Clarified Activated Sludge with Nitrifying Trickling Filter Biomass (20.6°C, 7.5 S.U.).

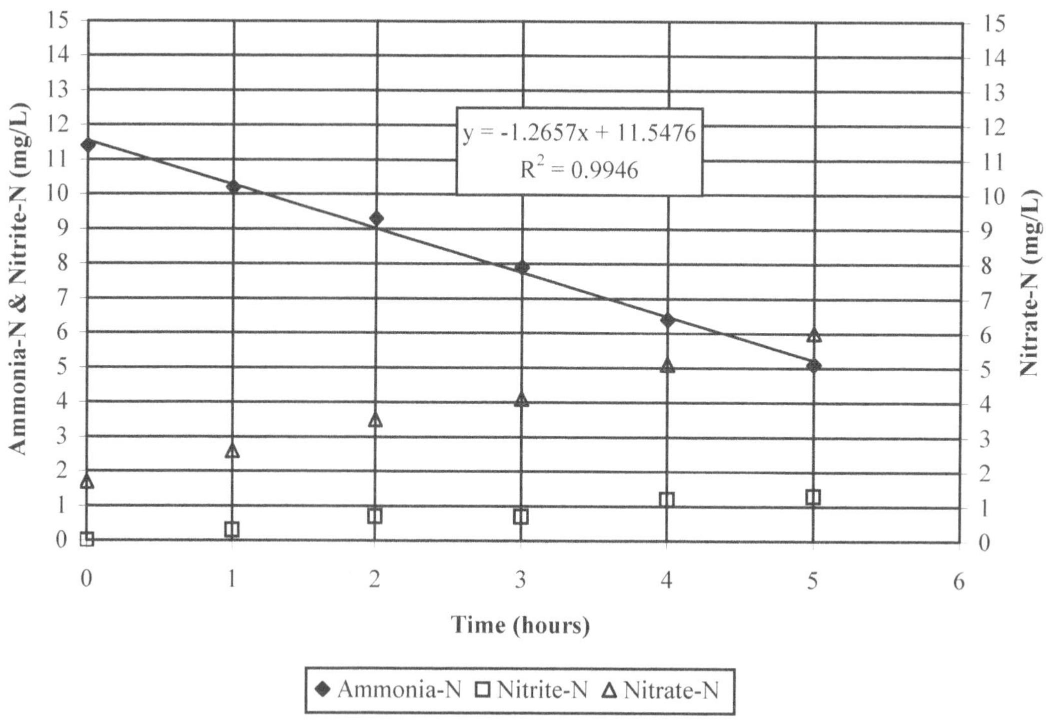

Figure 4-21. Run #17 (Duplicate); Bioaugmentation in Clarified Activated Sludge with Nitrifying Trickling Filter Biomass (20.9°C, 7.5 S.U.).

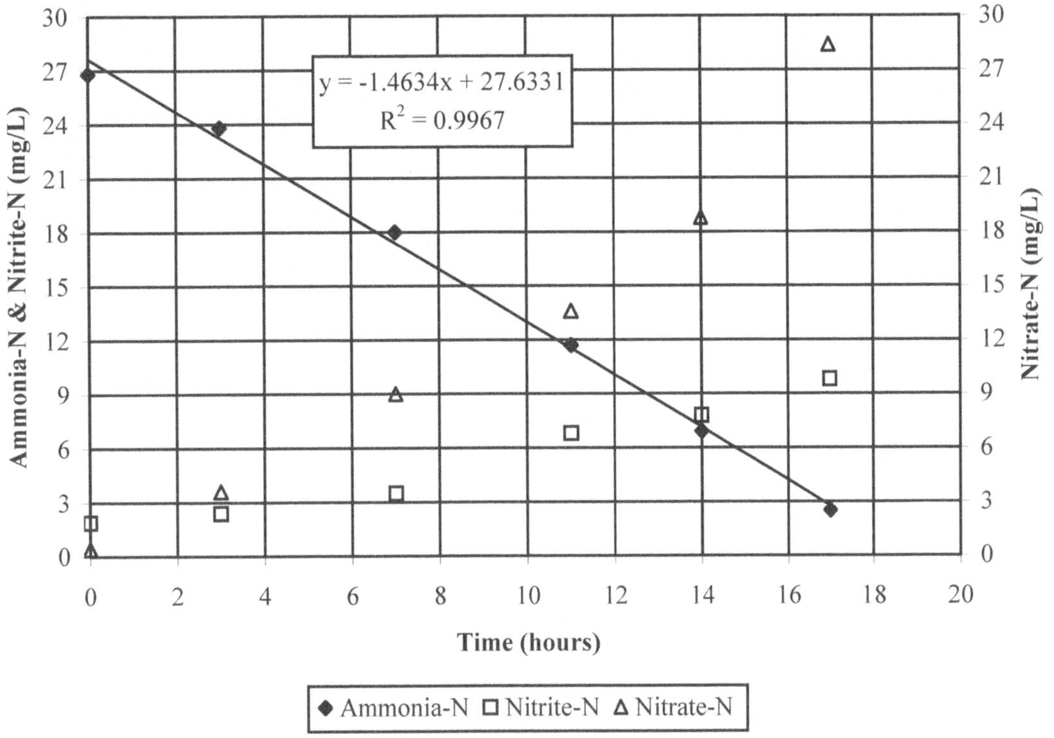

Figure 4-22. Run #18; Bioaugmentation in Activated Sludge with Nitrifying Trickling Filter Biomass (21.7°C, 7.4 S.U.).

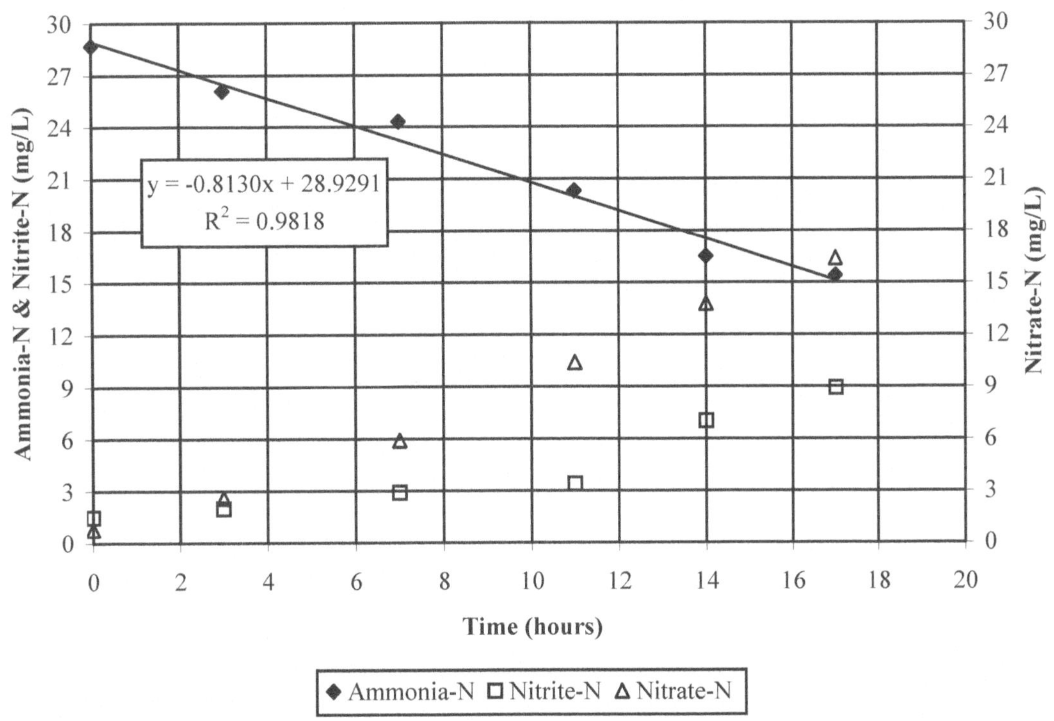

Figure 4-23. Run #19; Bioaugmentation in Activated Sludge with Nitrifying Trickling Filter Biomass (21.3°C, 7.5 S.U.).

The observed ammonium oxidation rates exhibited very little variation between substrate types; 0.066–0.084 mg NH_3-N oxidized/mg VSS per day for all test runs. The average observed ammonium oxidation rates were 0.075 mg NH_3-N oxidized/mg VSS per day for the low-strength ammonia synthetic wastewater, 0.078 mg NH_3-N oxidized/mg VSS per day for the clarified activated sludge, and 0.071 NH_3-N oxidized/mg VSS per day for the activated sludge.

The results of the bioaugmentation analyses indicate that acclimation was not a critical consideration for bioaugmentation with nitrifying trickling filter biomass in a nonstressed environment.

CHAPTER 5.0

CONCLUSIONS AND RECOMMENDATIONS

5.1 Development of Cultures of Nitrifiers from High-strength Ammonia Wastewater for Subsequent Bioaugmentation

Nitrification kinetics were evaluated in bench-scale batch reactors fed with high-strength ammonia synthetic wastewater (~1000 mg NH_3-N/L) operated at SRTs of 5, 10, and 20 days and with dewatered biosolids supernatant (1126–1680 mg NH_3-N/L) operated at an SRT of 20 days.

For the synthetic wastewater reactor operated at a 5-day SRT, partial conversion of ammonia to nitrite was observed with little or no conversion of nitrite to nitrate. The quasi-steady-state ammonium oxidation rate was 1.51 g NH_3-N oxidized/g VSS_N per day with a yield coefficient of 0.13 g VSS_N/g NH_3-N removed. At this SRT, un-ionized ammonia appeared to inhibit ammonium oxidation and prevent the development of an active nitrite oxidizing population.

For the synthetic wastewater reactor operated at a 10-day SRT, cycling between ammonia removal, nitrite production and removal, and nitrate production was observed. This cycling appeared to be related to inhibitory concentrations of un-ionized ammonia and un-ionized nitrous acid. The data indicate a sensitive balance between ammonia oxidizing microorganisms and nitrite oxidizing microorganisms where changes in the concentrations of these constituents directly affect the predominant biological population and activity. When conditions near inhibitory levels were experienced, established or acclimated populations continued to function beyond inhibitory levels to a certain degree. Similarly, when conditions changed from inhibitory to noninhibitory, a lag was experienced during which a new population of microorganisms became established. The quasi-steady-state ammonium oxidation rate (0.86 g NH_3-N oxidized/g VSS_N per day) was lower than that for the 5-day SRT while the yield coefficient was approximately the same (0.12 g VSS_N/g NH_3-N removed). Although a greater degree of nitrification occurred, steady-state operation was not observed.

For the synthetic wastewater reactor operated at a 20-day SRT, near complete conversion of ammonia to nitrate was observed. A steady-state condition without inhibitory conditions was achieved. The steady-state ammonium oxidation rate (0.33 g NH_3-N oxidized/g VSS_N per day) was lower than the 5- and 10-day SRT reactors with a comparable yield coefficient (0.15 g VSS_N/g NH_3-N removed).

Near complete conversion of ammonia to nitrate was also observed in the dewatered biosolids supernatant reactor operating at a 20-day SRT without the occurrence of inhibitory concentrations of un-ionized ammonia or un-ionized nitrous acid. The ammonium oxidation rate was 0.21 g NH_3-N oxidized/g VSS_N per day with a yield coefficient of 0.24 g VSS_N/g NH_3-N removed.

Near complete nitrification with conversion to nitrate was achieved at higher SRT values exhibiting steady state operation. However, steady-state operation was also observed to depend on the concentrations of ammonia, un-ionized ammonia, and/or un-ionized nitrous acid that were affected by the start-up and operational conditions associated with each SRT (in particular, start-up with high-strength ammonia waste and batch feed operation). Lower SRTs may be able to achieve stable, near complete nitrification under different start-up and operational conditions.

Although it may be possible to culture a population capable of near complete nitrification at a lower SRT, the design configuration and operational strategy must consider, mitigate, and/or eliminate the potential for un-ionized ammonia and un-ionized nitrous acid inhibition (e.g., process start-up at lower concentration with gradual increase to higher concentration, continuous feed). The operation of a full-scale, low-SRT system at steady state may also be very difficult from a practical perspective.

5.2 Bioaugmentation in Non-nitrifying Activated Sludge

Batch bioaugmentation analyses were conducted in the MLSS and final clarifier effluent from a non-nitrifying activated sludge with seed nitrifiers developed from the high-strength ammonia synthetic wastewater and dewatered biosolids supernatant operated at a 20-day SRT and also with biomass from an operating nitrifying trickling filter facility. Environmental conditions (e.g., temperature and pH) between the seed bioreactors (used to develop the supplemental nitrifiers) and the bioaugmentation reactors (to which supplemental nitrifiers were added) were not significantly different.

Ammonia removal was observed in all bioaugmentation analyses with no apparent lag or acclimation period. Nitrite production was also immediately observed. In all of the bioaugmentation analyses with nitrifiers developed from high-strength ammonia wastewater and dewatered biosolids supernatant, ammonia was initially converted to nitrite with little, if any, conversion to nitrate. It would appear that conditions more favorable to ammonia oxidizing microorganisms exist initially and that, as the ammonia concentration is reduced, the oxidation of nitrite to nitrate can begin to take place. Although nitrite accumulation was observed to some degree in all test runs, the greatest accumulation (based on the maximum nitrite concentration measured) occurred in the analyses with MLSS.

Acclimation was evaluated by comparing the initial ammonia removal, nitrite production, and nitrate production. Kinetic parameters from the seed bioreactors used to develop the supplemental nitrifiers were also compared to kinetic parameters from the bioaugmentation analyses. Observed and/or estimated ammonium oxidation rates were not significantly different between the seed reactors and the batch bioaugmentation analyses. Therefore, acclimation does not appear to be a critical obstacle for nitrifier bioaugmentation in conventional activated sludge systems when environmental conditions between the seed and bioaugmentation processes are not significantly different.

Nitrate production occurred almost immediately with the trickling filter biomass. Observed ammonium oxidation rates for the bioaugmentation analyses with nitrifying trickling filter biomass were significantly less than those from the dewatered biosolids supernatant seed reactor indicating that the trickling filter biomass contained a lower fraction of active nitrifiers.

5.3 Recommendations

Bioaugmentation in conventional activated sludge systems with nitrifiers developed from high-strength ammonia wastewater sidestream treatment processes does not appear to be significantly complicated by acclimation issues when environmental conditions (e.g., temperature, pH, DO) are controlled. Therefore, it appears that there is sufficient technical basis to conduct additional study of this concept. A continuous-flow pilot study of the complete system, including a side-stream treatment system for developing nitrifiers from high-strength ammonia wastewater and a mainstream non-nitrifying activated sludge process, is recommended. In particular further study should consider the optimal side-stream system design, start-up, and operation at an appropriate SRT to provide a yield of nitrifiers sufficient to achieve near-complete nitrification in the mainstream activated sludge process.

REFERENCES

Abeysinghe, D.H.; De Silva, D.G.V.; Stahl, D.A.; Rittmann, B.E. The effectiveness of bioaugmentation in nitrifying systems stressed by a washout condition and cold temperature. *Water Environ. Res.* **2002,** *74,*187–199.

American Public Health Association. *Standard Methods for the Examination of Water and Wastewater*, 20th ed. American Public Health Association: Washington, D.C., 1998.

Anthonisen, A.C.; Loehr, R.C.; Prakasam, T.B.S., Srinath, E.G. Inhibition of nitrification by ammonia and nitrous acid. *J. Water Pollution Control Fed.* **1976**, *48*, 835–852.

Constantine, T.A.; Filipe, C.D.M.; Elliott, M.D.; Crawford, G.V.; Daigger, G.T. New nitrifier bioaugmentation process configuration to achieve year round nitrification at low SRTs. In *Proceedings of the 74th Water Environment Federation Annual Conference,* Atlanta, GA, Oct 13–17, 2001; Water Environment Federation, Alexandria, VA, 2001.

De Silva, D.G.V.; Abeysinghe, D.H.; Rittmann, B.E. Effect of bioaugmentation in stressed wastewater treatment systems. In *Proceedings of the 73rd Water Environment Federation Annual Conference,* Oct 14–18, 2000, Anaheim, CA; Water Environment Federation, Alexandria, VA, 2000.

Head, M.A.; Oleszkiewicz, J.A. Short SRT nitrification through seeding nitrifying bacteria into cold, non-nitrifying sequencing batch reactors. In *Proceedings of the 75th Water Environment Federation Annual Conference*, Chicago, Sept 28–Oct 2, 2002; Water Environment Federation, Alexandria, VA, 2002.

Kos, P.; Head, M.A.; Oleszkiewicz, J.; Warakomski, A. Demonstration of low temperature nitrification with a short SRT. In *Proceedings of the 73rd Water Environment Federation Annual Conference,* Anaheim, CA, Oct 14–18, 2000; Water Environment Federation, Alexandria, VA; 2000.

Kos, P. Short SRT (solids retention time) nitrification process/flowsheet. *Water Sci. Technol.* **1998**, *38*, 23–29.

Mossakowska, A.; Reinius, L.-G.; Hultman, B. Nitrification reactions in treatment of supernatant from dewatering of digested sludge. *Water Environ. Res.* **1997**, *69*, 1128–1133.

Noto, K.; Ogasawara, T.; Suwa, Y.; Sumino, T. Complete oxidation of high concentration of ammonia by retaining incompatible nitrification activities in three-vessel system. *Water Res.* **1998**, *32*, 769–773.

Sedlak, R.I. *Phosphorus and Nitrogen Removal from Municipal Wastewaters: Principles and Practice,* 2nd ed. Lewis Publishers: Ann Arbor, MI, 1991.

Soriano, A.G.; Erb, M.; Garel, C.; Audic, J.M. A comparative pilot-scale study of the performance of conventional activated sludge and membrane bioreactors under limiting operating conditions. *Water Environ. Res.* **2003**, *75,* 225–231.

Stensel, H.D.; Ke, D.L.; Dold, P.L.; Jones, R.; Bye, C.; Melcer, H. (2002) Methods for measuring nitrifier maximum specific growth rate: Crucial for plant capacity rating. In

Proceedings of the 75th Water Environment Federation Annual Conference, Chicago, Sept 28–Oct 2, 2002; Water Environment Federation, Alexandria, VA, 2002.

Surmacz-Gorska, J.; Andrzej Cichon, A.; Miksch, K. Nitrogen removal from wastewater with high ammonia nitrogen concentrations via shorter nitrification and denitrification. *Water Sci. Technol.* **1997**, *36*, 73–78.

U.S. Environmental Protection Agency. *Nitrogen Control Manual,* EPA/625/R-93/010. Office of Research and Development, U.S. Environmental Protection Agency: Washington, D.C., 1993.

WERF SUBSCRIBERS

WASTEWATER UTILITY

Alabama
Montgomery Water Works & Sanitary Sewer Board

Alaska
Anchorage Water & Wastewater Utility

Arizona
Gila Resources

Glendale, City of, Utilities Department

Mesa, City of

Peoria, City of

Phoenix Water Services Department

Pima County Wastewater Management

Arkansas
Little Rock Wastewater Utility

California
Calaveras County Water District

Central Contra Costa Sanitary District

Contra Costa Water District

Crestline Sanitation District

Delta Diablo Sanitation District

Dublin San Ramon Services District

East Bay Dischargers Authority

East Bay Municipal Utility District

El Dorado Irrigation District

Fairfield-Suisun Sewer District

Fresno Department of Public Utilities

Irvine Ranch Water District

Las Virgenes Municipal Water District

Livermore, City of

Lodi, City of

Los Angeles, City of

Los Angeles County, Sanitation Districts of

Napa Sanitation District

Orange County Sanitation District

Palo Alto, City of

Riverside, City of

Sacramento Regional County Sanitation District

San Diego Metropolitan Wastewater Department, City of

San Francisco, City & County of

San Jose, City of

Santa Barbara, City of

Santa Rosa, City of

South Bayside System Authority

South Coast Water District

South Orange County Wastewater Authority

Union Sanitary District

West Valley Sanitation District

Colorado
Boulder, City of

Colorado Springs Utilities

Littleton/Englewood Water Pollution Control Plant

Metro Wastewater Reclamation District, Denver

Connecticut
The Mattabasset District

New Haven, City of, WPCA

District of Columbia
District of Columbia Water & Sewer Authority

Florida
Broward, County of

Fort Lauderdale, City of

Gainesville Regional Utilities

JEA

Miami-Dade Water & Sewer Authority

Orange County Utilities Department

Orlando, City of

Reedy Creek Improvement District

Seminole County Environmental Services

St. Petersburg, City of

Stuart Public Utilities

Tallahassee, City of

Tampa, City of

Toho Water Authority

West Palm Beach, City of

Georgia
Atlanta Department of Watershed Management

Augusta, City of

Clayton County Water Authority

Cobb County Water System

Columbus Water Works

Fulton County

Gwinnett County Department of Public Utilities

Macon Water Authority

Savannah, City of

Hawaii
Honolulu, City and County of

Illinois
American Bottoms Wastewater Treatment Plant

Greater Peoria Sanitary District

Kankakee River Metropolitan Agency

Metropolitan Water Reclamation District of Greater Chicago

North Shore Sanitary District

Wheaton Sanitary District

Indiana
Fort Wayne, City of

Iowa
Ames, City of

Cedar Rapids Wastewater Facility

Des Moines Metro Wastewater Reclamation Authority

Iowa City, City of

Kansas
Johnson County Unified Wastewater Districts

Unified Government of Wyandotte County/Kansas City, City of

Kentucky
Louisville & Jefferson County Metropolitan Sewer District

Louisiana
Sewerage & Water Board of New Orleans

Maine
Bangor, City of

Portland Water District

Maryland
Anne Arundel County Bureau of Utility Operations

Howard County Department of Public Works

Washington Suburban Sanitary Commission

Massachusetts
Boston Water & Sewer Commission

Upper Blackstone Water Pollution Abatement District

Michigan
Ann Arbor, City of

Detroit, City of

Holland Board of Public Works

Lansing, City of

Owosso Mid-Shiawassee County WWTP

Saginaw, City of

Wayne County Department of Environment

Wyoming, City of

Minnesota
Rochester, City of

Western Lake Superior Sanitary District

Missouri
Independence, City of

Kansas City Missouri Water Services Department

Little Blue Valley Sewer District

Nebraska
Lincoln Wastewater System

Nevada
Henderson, City of

New Jersey
Passaic Valley Sewerage Commissioners

New York
New York City Department of Environmental Protection

Rockland County Solid Waste Management Authority/ Sewer District

North Carolina
Cary, Town of

Charlotte/Mecklenburg Utilities

Durham, City of

Metropolitan Sewerage District of Buncombe County

Orange Water & Sewer Authority

Ohio
Akron, City of

Butler County Department of Environmental Services

Columbus, City of

Metropolitan Sewer District of Greater Cincinnati

Northeast Ohio Regional Sewer District

Summit, County of

Oklahoma
Tulsa, City of

Oregon
Clean Water Services

Eugene/Springfield Water Pollution Control

Water Environment Services

Pennsylvania
Philadelphia, City of

University Area Joint Authority, State College

South Carolina
Charleston Commissioners of Public Works

Mount Pleasant Waterworks & Sewer Commission

Spartanburg Sanitary Sewer District

Tennessee
Cleveland, City of

Knoxville Utilities Board

Murfreesboro Water & Sewer Department

Nashville Metro Water Services

Texas
Austin, City of

Dallas Water Utilities

Denton, City of

El Paso Water Utilities

Fort Worth, City of

Gulf Coast Waste Disposal Authority

WERF SUBSCRIBERS

Houston, City of

San Antonio Water System

Trinity River Authority

Utah

Salt Lake City Corporation

Virginia

Alexandria Sanitation Authority

Arlington, County of

Fairfax County Virginia

Hampton Roads Sanitation District

Henrico, County of

Hopewell Regional Wastewater Treatment Facility

Loudoun County Sanitation Authority

Lynchburg Regional WWTP

Prince William County Service Authority

Richmond, City of

Rivanna Water & Sewer Authority

Washington

Edmonds, City of

Everett, City of

King County Department of Natural Resources

Seattle Public Utilities

Sunnyside, Port of

Yakima, City of

Wisconsin

Green Bay Metro Sewerage District

Madison Metropolitan Sewerage District

Milwaukee Metropolitan Sewerage District

Racine, City of

Sheboygan Regional Wastewater Treatment

Wausau Water Works

Australia

South Australian Water Corp.

Sydney Water Corp.

Water Corp. of Western Australia

Canada

Greater Vancouver Regional District

Toronto, City of, Ontario

Winnipeg, City of, Manitoba

Mexico

Servicios de Agua y Drenaje de Monterrey, I.P.D.

New Zealand

Watercare Services Limited

Singapore

Singapore Public Utilities Board

United Kingdom

Yorkshire Water Services Limited

STORMWATER UTILITY

California

Los Angeles, City of, Department of Public Works

Monterey, City of

Santa Rosa, City of

Colorado

Boulder, City of

Georgia

Griffin, City of

Iowa

Cedar Rapids Wastewater Facility

Des Moines Metro Wastewater Reclamation Authority

Kansas

Overland Park, City of

Kentucky

Louisville & Jefferson County Metropolitan Sewer District

Maine

Portland Water District

Minnesota

Western Lake Superior Sanitary District

North Carolina

Cary, Town of

Charlotte, City of, Stormwater Services

Pennsylvania

Philadelphia, City of

Tennessee

Chattanooga Storm Water Management

Washington

Seattle Public Utilities

STATE

Arkansas Department of Environmental Quality

Fresno Metropolitan Flood Control District, Calif.

Urban Drainage & Flood Control District, Colo.

CORPORATE

ADS Environmental Services

The ADVENT Group Inc.

Alan Plummer & Associates

Alden Research Laboratory

Alpine Technology Inc.

Aqua-Aerobic Systems, Inc.

Aquateam–Norwegian Water Technology Centre A/S

BaySaver Inc.

BioVir Laboratories, Inc.

Black & Veatch

Boyle Engineering Corporation

BPR CSO

Brown & Caldwell

Burns & McDonnell

CABE Associates Inc.

The Cadmus Group

Camp Dresser & McKee Inc.

Carollo Engineers Inc.

Carpenter Environmental Associates Inc.

CDS Technologies Inc.

Chemtrac Systems Inc.

CH2M HILL

Damon S. Williams Associates, LLC

David L. Sheridan, P.C.

Earth Tech Inc.

Eco-Matrix

Environmental Engineers International

EMA Inc.

The Eshelman Company Inc.

Finkbeiner, Pettis, & Strout Inc.

Freese & Nichols, Inc.

Frontier Geosciences Inc.

ftn Associates Inc.

Fuss & O'Neill, Inc.

Gannett Fleming Inc.

Golder Associates Inc.

Greeley and Hansen LLC

Hazen & Sawyer, P.C.

HDR Engineering Inc.

HNTB Corporation

HydroQual Inc.

Ingersoll-Rand Energy Systems

Insituform Technologies Inc.

Institute for Environmental Technology & Industry, Korea

Jacobson Helgoth Consultants Inc.

Jason Consultants Inc.

Jordan, Jones, & Goulding Inc.

KCI Technologies Inc.

Kelly & Weaver, P.C.

Kennedy/Jenks Consultants

Komline Sanderson Engineering Corporation

Lawler, Matusky & Skelly Engineers, LLP

Limno-Tech Inc.

Lombardo Associates Inc.

Malcolm Pirnie Inc.

McKim & Creed

MEC Analytical Systems Inc.

Metcalf & Eddy Inc.

MWH

New England Organics

Odor & Corrosion Technology Consultants Inc. (OCTC)

ONDEO Degremont Inc.

Oswald Green, LLC

PA Government Services Inc.

Parametrix Inc.

Parsons

Post, Buckley, Schuh & Jernigan

R&D Engineering/Conestoga Rover & Associates

The RETEC Group

RMC, Inc. (Raines, Melton & Carella)

R.M. Towill Corporation

Ross & Associates Ltd.

Royce Technologies

SAIC Maritime Technical Group

Sear-Brown

Short Elliott Hendrickson, Inc.

Stormwater Management, Inc.

Synagro Technologies, Inc.

Tetra Tech Inc.

Trojan Technologies Inc.

URS Corporation

Veolia Water NATC

Wade-Trim Inc.

Weston Solutions Inc.

Woodard & Curran

WRc/D&B, LLC

WWETCO, LLC

Zoeller Pump Company

INDUSTRY

American Electric Power

ChevronTexaco Energy Research & Technology Company

The Coca-Cola Company

Dow Chemical Company

DuPont Company

Eastman Kodak Company

Merck & Company Inc.

ONDEO Services

Procter & Gamble Company

PSEG Services Corp.

Severn Trent Services Inc.

Shell Global Solutions

Thames Water Plc

United Water Services LLC

Note: List as of 1/31/04

WERF Product Order Form

As a benefit of joining the Water Environment Research Foundation, subscribers are entitled to receive one complimentary copy of all final reports and other products.Additional copies are available at cost (usually $10).To order your complimentary copy of a report, please write "free" in the unit price column.WERF keeps track of all orders.If the charge differs from what is shown here, we will call to confirm the total before processing.

Name ____________________ Title ____________________

Organization ____________________

Address ____________________

City __________ State __________ Zip Code __________ Country __________

Phone __________ Fax __________ Email __________

Stock #	Product	Quantity	Unit Price	Total
			Postage & Handling	
			VA Residents Add 4.5% Sales Tax	
			Canadian Residents Add 7% GST	
			TOTAL	

Method of Payment: (All orders must be prepaid.)

❑ Check or Money Order Enclosed

❑ Visa ❑ Mastercard ❑ American Express

Account No. ____________________ Exp. Date __________

Signature ____________________

Shipping & Handling:

Amount of Order	United States	Canada & Mexico	All Others
Up to but not more than:	Add:	Add:	Add:
$20.00	$5.00*	$8.00	50% of amount
30.00	5.50	8.00	40% of amount
40.00	6.00	8.00	
50.00	6.50	14.00	
60.00	7.00	14.00	
80.00	8.00	14.00	
100.00	10.00	21.00	
150.00	12.50	28.00	
200.00	15.00	35.00	
More than $200.00	Add 20% of order	Add 20% of order	

*minimum amount for all orders

Note: Please make checks payable to the Water Environment Research Foundation.

To Order (Subscribers Only):

Log on to www.werf.org and click on the "Product Catalog."

Phone: (703) 684-2470
Fax: (703) 299-0742.

WERF
Attn: Subscriber Services
635 Slaters Lane
Alexandria, VA 22314-1177

To Order (Non-Subscribers):

Non-subscribers may be able to order WERF publications either through WEF (www.wef.org) or IWAP (www.iwapublishing.com).Visit WERF's website at www.werf.org for details.